CONTENTS

W9-BUX-473

 **PART I** **PREPARATION AND SUPPORT**

 **PART II** **ACTIVITIES AND LESSON NOTES**

 **PART III** **SUPPLEMENTARY CUTOUTS**

A TOPS Teaching Model

If science were only a set of explanations and a collection of facts, you could teach it with blackboard and chalk. You could require students to read chapters in a textbook, assign questions at the end of each chapter, and set periodic written exams to determine what they remember. Science is traditionally taught in this manner. Everybody studies the same information at the same time. Class togetherness is preserved.

But science is more than this. It is also process — a dynamic interaction of rational inquiry and creative play. Scientists probe, poke, handle, observe, question, think up theories, test ideas, jump to conclusions, make mistakes, revise, synthesize, communicate, disagree and discover. Students can understand science as process only if they are free to think and act like scientists, in a classroom that recognizes and honors individual differences.

Science is *both* a traditional body of knowledge *and* an individualized process of creative inquiry. Science as process cannot ignore tradition. We stand on the shoulders of those who have gone before. If each generation reinvents the wheel, there is no time to discover the stars. Nor can traditional science continue to evolve and redefine itself without process. Science without this cutting edge of discovery is a static, dead thing.

Here is a teaching model that combines both the content and process of science into an integrated whole. This model, like any scientific theory, must give way over time to new and better ideas. We challenge you to incorporate this TOPS model into your own teaching practice. Change it and make it better so it works for *you*.

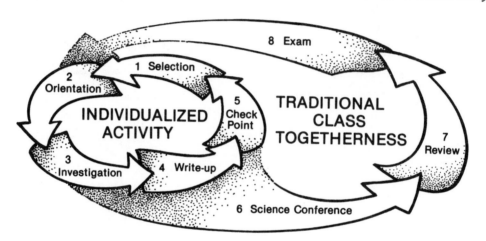

1. SELECTION

Students generally select worksheets in sequence, because new concepts build on old ones in a specific order. There are, however, exceptions to this pattern: students might *skip* a lesson that is not challenging; *repeat* a lesson with doubtful results; *add* an experiment to answer their own "what-would-happen-if?" questions.

Working at their own pace, students fall into a natural routine that creates stability and order. They still have questions and problems, to be sure, but remain purposefully engaged with a definite sense of direction.

2. ORIENTATION

Any student with basic reading skills can successfully interpret our carefully designed worksheet directions. If your class is new to TOPS, it may take a while for your students to get used to following directions by themselves, and to trust in their own problem-solving ability.

When students ask you for help, first ask them to read what they don't understand. If they didn't read the instruction in the first place, this should clear things up. Identify poor readers in your class. When they ask, "What does this mean?" they may be asking in reality, "Will you please read these directions aloud?"

Beyond reading comprehension, certain basic concepts and skills are also necessary to complete many worksheets. You can't, for example, ask students to measure something unless they know how to use a ruler. *Supporting Concepts* in our Teaching Notes list these basics plus strategies for teaching them. Teach these prerequisites at the beginning of class periods, before students begin their daily individualized work. Anticipate the needs and abilities of your particular age group: primary students will need more introductory support than middle school students; secondary students may require none at all.

A

THE EARTH, MOON & SUN

with paper plates,
bottles,
tennis balls
and simple things

SCIENCE WITH SIMPLE THINGS SERIES

Conceived and
written by

RON MARSON

Illustrated by

PEG MARSON

 LEARNING
SYSTEMS

10970 S. Mulino Rd.
Canby, Oregon 97013

ISBN 0-941008-40-1

Printed on Recycled Paper ♻

3. INVESTIGATION

Students work through the worksheets independently and cooperatively. They follow their own experimental strategies and help each other. You can encourage this behavior by helping students only *after* they have tried to help themselves. As a resource teacher, you work to stay *out* of the center of attention, responding to student questions rather than posing teacher questions.

Some students will progress more rapidly through these worksheets than others. To finish as a cohesive group, announce well in advance when individualized worksheet activity will end. Expect to generate a frenzy of activity as students rush to meet your deadline. While slower students finish the core activities you specify, challenge your more advanced students with open-ended *Extension* activities from our teaching notes, or to design their own original experiments.

4. WRITE-UP

Underlined numbers in the worksheets signal that students should explain the how and why of things with answers written on their own papers. Answers may be brief and to the point, with the exception of those that require creative writing. Students may accelerate their pace by completing these reports out of class.

Students may work alone, or in cooperative lab groups. But each one should prepare an original write-up, and bring it to you for approval. Avoid an avalanche of write-ups near the end of the module by enforcing this simple rule: each write-up must be approved *before* starting the next activity.

5. CHECK POINT

Student and teacher together evaluate each write-up on a pass/no-pass basis. Thus no time is wasted haggling over grades. If the student has made reasonable effort consistent with individual ability, check off the completed activity on a progress chart. Students keep these in notebooks or assignment folders kept on file in class.

Because the student is present when you evaluate, feedback is immediate and effective. A few moments of your personal attention is surely more effective than tedious margin notes that students may not heed or understand. Remember, you don't have to point out every error. Zero in on particular weaknesses. If reasonable effort is not evident, direct students to make specific improvements and return for a final check.

A responsible lab assistant can double the amount of individual attention each student receives. If he

or she is mature and respected by your students, have the assistant check even-numbered reports, while you check the odd ones. This will balance the work load and assure that everyone receives equal treatment.

6. SCIENCE CONFERENCE

Individualized worksheet activity has ended. This is a time for students to come together, to discuss experimental results, to debate and draw conclusions. Slower students learn about the enrichment activities of faster classmates. Those who did original investigations or made unusual discoveries share this information with their peers, just like scientists at a real conference.

This conference is an opportunity to expand ideas, explore relevancy, and integrate subject areas. Consider bringing in films, newspaper articles and community speakers. It's a meaningful time to investigate the technological and social implications of the topic you are studying. Make it an event!

7. REVIEW

Does your school have an adopted science textbook? Do parts of your science syllabus still need to be covered? Now is the time to integrate traditional science resources into your overall program. Your students already share a common background of hands-on lab work. With this base of experience, they can now read the text with greater understanding, think and problem-solve more successfully, communicate more effectively.

You might spend just a day here, or an entire week. Finish with a review of major concepts in preparation for the final exam. Our review/test questions provide an excellent basis for discussion and study.

8. EXAM

Use any combination of our review/test questions, plus questions of your own, to determine how well students have mastered the concepts they've been learning. Those who finish your exam early might be eager to begin work in the next new TOPS module.

Now that your class has completed a major TOPS learning cycle, it's time to start fresh with a brand new topic. Those who messed up and got behind don't need to stay there. Everyone begins the new topic on an equal footing. This frequent change of pace encourages your students to work hard, to enjoy what they learn, and thereby grow in scientific literacy.

B

Getting Ready

Here is a checklist of things to think about and preparations to make before beginning your first lesson on THE EARTH, MOON AND SUN.

✔ Review the scope and sequence.

Take just a few minutes, right now, to page through all 20 lessons. Pause to read each *Objective* (top left corner of the Teaching Notes) and scan each lesson.

✔ Set aside appropriate class time.

Allow an average of perhaps 1 class period per lesson (more for younger students), plus time at the end of this module for discussion, review and testing. If you teach science every day, this module will likely engage your class for about 5 weeks. If your schedule doesn't allow this much science, consult the logic tree on page E to see which activities you can safely omit without breaking conceptual links between lessons.

✔ Decide when to start.

Teach this daytime astronomy module during any season of the year. An optimal time to begin is at the new moon, or perhaps a day or two earlier, but other times of the month are OK too. Keep in mind that 6 out of 20 lessons require direct sunlight. If the sun doesn't cooperate, you can easily skip ahead, then return when the weather clears.

✔ Number your activity sheet masters.

The small number printed in the top right corner of each activity sheet shows its position within the series. If this ordering fits your schedule, copy each number into the blank parentheses next to it. Be sure to use pencil; you may decide to revise, rearrange, add or omit lessons the next time you teach this module. Insert your own better ideas wherever they fit best, and renumber the sequence. This allows your curriculum to adapt and grow as you do.

✔ Photocopy sets of student activity sheets.

All activity sheets in this module can be reused. Photocopy and collate a classroom set to use year after year, as you would any textbook. All questions emphasized with underlined numbers require students to respond on separate paper. Find all supplementary materials that support each lesson on reproducible pages at the back of this module. The *Materials* list that accompanies each lesson tells you when these cutouts are needed, and how many to photocopy.

Please observe our copyright notice at the front of this module. We allow you, the purchaser, to make as many copies as you need, but forbid supplying your materials to other teachers for use in other classrooms. Our only income is from the sale of these inexpensive modules. If you would like to help spread the word that TOPS is tops, please request multiple copies of our TOPS Ideas magazine/catalog (sent to you free and postpaid) to distribute to other faculty members or student teachers. These offer a variety of sample lessons and an order form, so your colleagues can purchase their own TOPS modules.

✔ Collect needed materials.

See page D for details.

✔ Organize a way to track assignments.

Keep student work on file in class. If you lack a file cabinet, a box with a brick will serve. File folders or notebooks both make suitable assignment organizers. Students will feel a sense of accomplishment as they see their folders grow heavy, or their notebooks fill, with completed assignments. Since all papers stay together, reference and review are easy.

Ask students to number a sheet of paper from 1 to 20 and tape it inside the front cover of their folders or notebooks. Track individual progress through this module (and future modules) by initialing lesson numbers as completed.

✔ Review safety procedures.

In our litigation-conscious society, we find that publishers are often more committed to protecting themselves from liability suits than protecting students from physical hazards. Lab instructions are often so filled with spurious advisories, cautions and warnings that students become desensitized to safety in general. If we cry "Wolf!" too often, real warnings of present danger may go unheeded.

At TOPS we endeavor to use good sense in deciding what students already know (don't stab yourself in the eye) and what they should be told (don't look directly at the sun.) Scissors and pins, of course, could be dangerous in the hands of unsupervised children. This curriculum cannot anticipate irresponsible behavior or negligence. It is ultimately the teacher's responsibility to see that common-sense safety rules are followed at all times. And it is your students' responsibility to respect and protect themselves and each other.

Unusual hazards detailed in this module are as follows:
• In activity 2, be careful not to poke your eye with the straw as you look through it.
• Never sight directly into the sun. Use shadows to find the sun's azimuth and altitude as detailed in activity 4.

✔ Communicate your grading expectations.

Whatever your grading philosophy, your students need to understand how they will be assessed. Here is a scheme that counts individual effort, attitude and overall achievement. We think these three components deserve equal weight:
• Pace (effort): Tally the number of check points and extra credit experiments you have initialed for each student. Low-ability students should be able to keep pace with gifted students, since write-ups are evaluated relative to individual performance standards on a pass/no-pass basis. Students with absences or those who tend to work slowly might assign themselves more homework out of class.
• Participation (attitude): This is a subjective grade, assigned to measure personal initiative and responsibility. Active participators who work to capacity receive high marks. Inactive onlookers who waste time in class and copy the results of others receive low marks.
• Exam (achievement): Activities point toward generalizations that provide a basis for hypothesizing and predicting. The test questions on pages G-J will help you assess whether students understand relevant theory and can apply it in a predictive way.

Gathering Materials

Listed below is *everything* you'll need to teach this module. You probably already have most items. Buy the rest locally, or ask students to bring recycled materials from home.

Keep this classification key in mind as you review what's needed:

general on-the-shelf materials:	special in-a-box materials:
Normal type suggests that these materials are used often. Keep these basics on shelves or in drawers that are accessible to your students. The next TOPS module you teach will likely utilize many of these same materials.	*Italic type suggests that these materials are unusual. Keep these specialty items in a separate box. After you finish teaching this module, label the box and put it away, ready to use again.*
(substituted materials):	*optional materials:
Parentheses enclosing any item suggest a ready substitute. These alternatives may work just as well as the original. Don't be afraid to improvise, to make do with what you have.	An asterisk sets these items apart. They are nice to have, but you can easily live without them. They are probably not worth an extra trip to the store, unless you are gathering other materials as well.

Everything is listed in order of first use. Start gathering at the top of this list and work down. The Teaching Notes may occasionally suggest additional *Extensions*. Materials for these optional experiments are listed neither here nor under *Materials*. Read the extension itself to determine what new items, if any, are required.

Needed quantities depend on how many students you have, how you organize them into activity groups, and how you teach. Decide which of these 3 estimates best applies to you, then adjust quantities up or down as necessary:

$Q_1/Q_2/Q_3$

Single Student: Enough for 1 student to do all the experiments.
Individualized Approach: Enough for 30 students informally working in pairs, all self-paced.
Traditional Approach: Enough for 30 students, organized into pairs, all doing the same lesson.

KEY:	general on-the-shelf materials (substituted materials)	*special in-a-box materials* *optional materials

$Q_1/Q_2/Q_3$

	general on-the-shelf		special in-a-box
1/1/1	pkt of steel pins, 1 inch long — see teaching notes 5	.5/7/7	quarts dry gravel or sand
1/1/1	spool of thread	3/3/3	full-sized newspaper sheets
1/15/15	scissors — high quality scissors recommended for activity 14	1/8/15	meter sticks
1/15/15	magnets — the ceramic magnets sold by TOPS are suitable	1/8/15	*hand calculators
		1/2/2	rolls adding machine tape
1/35/35	small baby food jars or equivalent	1/3/6	paper punch tools
1/1/1	water source — a large pitcher is suitable	1/5/15	U.S. nickels or equivalent-sized coin
3/35/35	straight plastic drinking straws — not wide milk shake straws	1/5/15	orange crayons or marking pens
.1/1/1	cup oil-based modeling clay	1/1/1	roll waxed paper
5/62/62	index cards, 4 x 6 inches	2/30/30	rubber bands — thick ones work best
1/10/15	rolls clear tape — best if you can write on it	1/1/1	aluminum foil
1/1/1	*roll clear tape, adhesive on both sides	1/1/1	roll kite string (heavy thread or dental floss)
1/10/15	rolls masking tape	1/8/15	clipboards (books)
1/1/1	pencil sharpener	2/30/30	*tennis balls, new or used*
3/75/75	generic paper plates — see notes 20	2/10/30	batteries, dead or alive — size D are best
1/30/30	medium-sized washers, 3/4 inch (19 mm) outside diameter	1/5/15	flashlights
1/8/15	textbooks	1/15/15	Ping-Pong balls
1/15/15	wristwatches	3/45/45	paper clips
7/60/110	medium cans of equal diameter (paper towel or gift wrap tubes) — see notes 10	1/1/1	*roll black tape — electrical, vinyl or cloth*
		3/45/45	*glass pop or beer bottles of equal height*
		1/5/15	cardboard milk cartons, quart or larger
		1/5/15	*canning rings, regular size
		1/1/1	*calendars for this year and probably the next — see notes 20*

Sequencing Activities

This logic tree shows how all the worksheets in this module tie together. In general, students begin at the trunk of the tree and work up through the related branches. Lower level activities support the ones above.

You may, at your discretion, omit certain activities or change their sequence to meet specific class needs. However, when leaves *open vertically* into each other, those below logically precede those above, and cannot be omitted.

When possible, students should complete the worksheets in the same sequence as numbered. If time is short, however, or certain students need to catch up, you can use this logic tree to identify concept-related *horizontal* activities. Some of these might be omitted since they serve to reinforce learned concepts rather than introduce new ones.

For whatever reason, when you wish to make sequence changes, you'll find this logic tree a valuable reference. Parentheses in the upper right corner of each worksheet allow you total flexibility. They are blank so you can pencil in sequence numbers of your own choosing.

Notice that 6 leaves on this tree are marked with an asterisk*. These activities can only be completed on *sunny* days. If the weather is cloudy, skip ahead, paying attention to how the lessons logically connect. Make up these worksheets when the weather clears.

THE EARTH, MOON & SUN 40

E

Long-Range Objectives

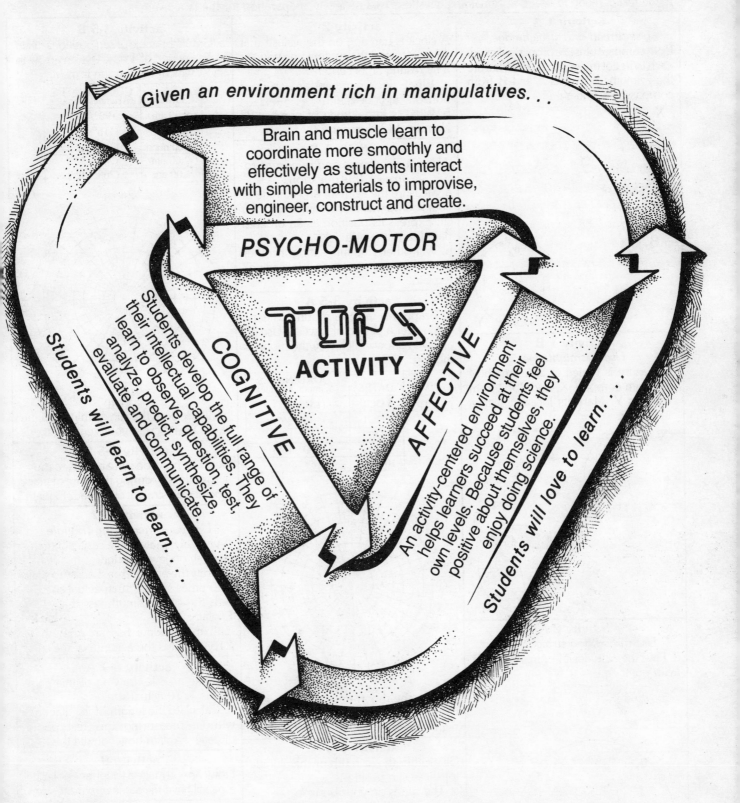

Given an environment rich in manipulatives. . .

Brain and muscle learn to coordinate more smoothly and effectively as students interact with simple materials to improvise, engineer, construct and create.

PSYCHO-MOTOR

TOPS ACTIVITY

COGNITIVE

Students develop the full range of their intellectual capabilities. They learn to observe, question, test, analyze, predict, synthesize, evaluate and communicate.

AFFECTIVE

An activity-centered environment helps learners succeed at their own levels. Because students feel positive about themselves, they enjoy doing science.

Students will learn to learn. . . .

Students will love to learn. . . .

Review / Test Questions

Photocopy both pages of test questions. On a separate sheet of blank paper, cut and paste those boxes you want to use. Include questions of your own design, as well. Crowd all these questions onto a single page for students to answer on another paper, or leave space for student responses after each question, as you wish. Duplicate a class set and your custom-made test is ready to use. Use remaining questions as a review in preparation for the final exam.

activity 1 A
Use thread or a straightedge.
Both compass pins are properly aligned, each to its dot (magnetic pole). What are the azimuths of points **A** and **B** from compasses X and Y?

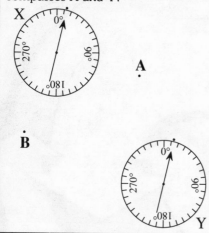

activity 1 B
Use a straightedge.
A treasure is buried at an azimuth of 78° from compass X, and 355° from compass Y. Mark its location with a circled dot.

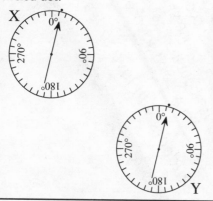

activity 2 A
Use thread or a straightedge.
a. Find the altitudes of points **A** and **B** from point X.

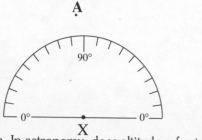

b. In astronomy, does altitude refer to height? Explain.

activity 2 B
A clock is hanging on the wall so that 12 points straight up. Imagine you are a fly resting at its center.
a. What clock numbers are on your horizon? at your zenith?
b. What is the altitude of number 2? Number 11? Number 4?

activity 1-3
Use your compass and quadrant to answer this question.
a. Measure the azimuth and altitude of your room's wall clock from where you sit. Express its position as an ordered pair: (azimuth, altitude)
b. Will other students sitting in other parts of the room get the same reading? Explain.

activity 3-5 A
a. Graph this table of moon positions taken over a period of 1 week. Use an arrow to show which way the moon is moving.

TIME	DATES	COORDINATES (Azimuth, Altitude)
7:30 pm	5 OCT	(151°, 27°)
	6 OCT	(136°, 26°)
	8 OCT	(111°, 16°)
	9 OCT	(100°, 15°)
	10 OCT	(89°, 6°)

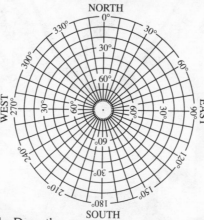

b. Does the moon appear to move in the direction of your arrow over 1 night? Explain.

activity 4-5 A
Give either the azimuth *or* the altitude for these sun positions:
a. The sun is on your horizon.
b. Your shadow has an azimuth of 40°.
c. The sun is shining, but you cast no shadow.
d. The sun culminates lower than your zenith.

activity 3-5 B
a. Graph this table of sun positions taken over a period of 1 day. Use an arrow to show which way the sun is moving.

TIME	COORDINATES (Azimuth, Altitude)
8:00 am	(92°, 32°)
10:00 am	(116°, 56°)
12:00 noon	(182°, 68°)
2:00 pm	(240°, 52°)
4:00 pm	(268°, 32°)

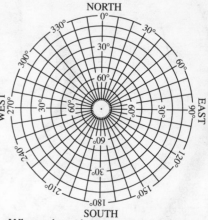

b. Where does the sun culminate in the sky? Does it cross your zenith?

activity 4-5 B
Describe how the sun appears to move across the sky during this time of year. Does it pass straight overhead? Explain.

activity 5-7
Use your measuring triangle.
A vertical 8 meter pole casts a 5 meter shadow on level ground.
a. Draw the pole and its shadow to scale 100 times smaller than actual size.
b. Is the sun's azimuth less than 45°? Explain.
c. If the shadow has an azimuth of 215°, what is the azimuth of the sun?

activity 6-7 A
a. Make a scale drawing of this page that is 1/10 actual size.
b. Calculate the length of its diagonal without measuring this test paper directly. Explain how you did this.

activity 8
How can you make a telephone pole and a pencil have the same *apparent* size?

activity 8-9 A
Using these numbers, write *four* true statements comparing the size of the earth, moon and sun and the distance between them: 3.5, 30, 108.

G

Answers

activity 1 A
From compass X: A is 104°, B is 202°.
From compass Y: A is 346°, B is 289°.

activity 1 B
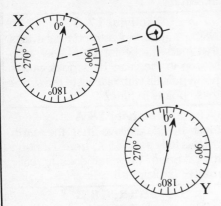

activity 2 A
a. From point X...
    the altitude of point A = 85°.
    the altitude of point B = 48°.
b. No. In astronomy, the altitude of an object refers to its angle with your horizon.

activity 2 B
a. The numbers 3 and 9 are on your horizon; 12 is at your zenith.
b. The number 2 has an altitude of 30°; eleven is at 60°; four lies below the horizon at –30°.

activity 1-3
a. (Answers will vary.)
b. No. The azimuth of any nearby point changes as you move around it; its altitude changes as you move closer or farther away.

activity 3-5 A
a.
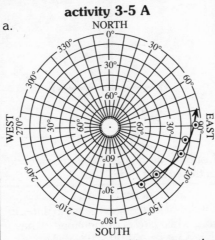
b. No. Because of earth's rotation, the moon appears to move from east to west, opposite its actual revolution around the earth from west to east.

activity 4-5 A
a. altitude = 0°   c. altitude = 90°
b. azimuth = 220°   d. azimuth = 180°

activity 3-5 B
a.
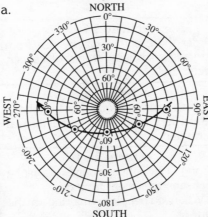
b. The sun culminates in the south, somewhat below the observer's zenith.

activity 4-5 B
(Various answers depending on season and latitude.) In general the sun rises in the east, culminates in the south, and sets in the west. Only in tropical latitudes does it pass straight overhead during certain seasons of the year.

activity 5-7
a. (This drawing is to scale.)
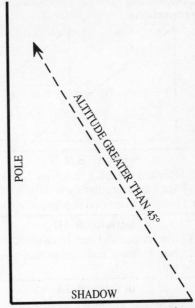
b. No. A 45° altitude is only halfway between your horizon and zenith. The shadow is considerably shorter than the pole, suggesting that the sun is higher than this.
c. Sun's azimuth = 215° – 180° = 35°

activity 6-7 A
a. (If your test paper measures a standard 8.5 x 11 inches, this converts to 21.6 x 27.9 cm.)

b. The diagonal across the scale drawing measures about 3.5 cm. Since the original paper is 10 times larger, its diagonal is about 35 cm.

activity 8
Hold a vertical pencil near enough to your eye so it appears to just cover a more distant telephone pole.

activity 8-9 A
3.5 moons fit across the earth.
108 earths fit across the sun.
30 earths reach from earth to the moon.
108 suns reach from the sun to the earth.

Review / Test Questions (continued)

activity 6-7 B

Use your measuring triangle.
Show that these triangles are proportional.

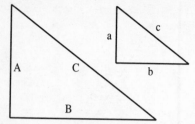

activity 8-9 B
a. How many earths fit across the sun?
b. How many moons fit across the earth?
a. How many moons fit across the sun?

activity 8-10
Do the moon and sun have the same apparent size? The same real size? Explain.

activity 8-11
Suppose the moon were reduced to 2 mm in diameter, about the size of a pinhead. At this scale…
a. Calculate the earth's diameter in mm.
b. Calculate the sun's diameter in cm.
c. Calculate the distance between them in meters.

activity 9
Relate the view of a basketball from a distance of 30 basketball diameters to a view of the earth and the moon.

activity 9-10 A
What apparent size would the earth and sun have as seen from the moon?

activity 9-10 B
To project a sun image of 1.0 cm in diameter, how long would you have to make your pinhole projector?

activity 10
How would you build a device to safely view an eclipse of the sun?

activity 10-11
From what distance should you view a basketball for it to have the same apparent size as the sun?

activity 12 A
a. If the moon were an umbrella and the earth were your head, how would you demonstrate an eclipse of the sun?
b. Is this a scale model?

activity 12 B
Half the earth is always in sunlight and half is in shadow.
a. Is this true for the moon as well?
b. If so, why does the moon sometimes appear as a thin crescent?

activity 13 A
A volleyball "moon" is balanced on a post, illuminated by a rising sun. Where should you stand to see a model of these moon phases?
  a. crescent
  b. gibbous
  c. quarter
  d. new
  e. full

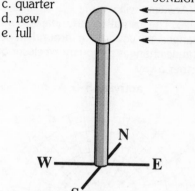

activity 13 B
List 8 moon phases in the moon cycle in the correct order.

activity 14 A
Circles of latitude on the earth differ in size, while circles of longitude all have the same size. Why is this so?

activity 14 B
Three principle circles divide a globe of the earth into equal halves:
a. Which is calibrated in degrees north and south?
b. Which is calibrated in degrees east and west?
a. Which is calibrated in months of the year?

activity 14 C
Order these earth belts by size from largest to smallest: tropic of cancer, arctic circle, equator.

activity 15 A
How should you align an earth globe to model the current alignment of the real earth and sun?

activity 15 B
A ring of twilight surrounds our globe that separates night from day.
a. Where do people on this ring see the sun?
b. Do you live on this ring?

activity 15-16
In which direction does the sun *appear* to move across the sky? Does it really move this way? Explain.

activity 16-17
What causes earth's changing seasons? Give an example.

activity 17 A
a. Where is the Tropic of Capricorn? Why is it important?
b. Where is the Arctic Circle? Why is it important?

activity 17 B
Name 4 important astronomical events that mark the beginning of each new season in the year. How does earth's north pole tilt with respect to the sun at these points in time?

activity 18 A
Most people believe that the earth rotates full circle (360°) from 12 noon to 12 noon. How far does it really rotate? Explain.

activity 18 B
Is 24 hours of star time the same as 24 hours of solar time? Explain.

activity 3, 19
Which way does the moon *appear* to move? Which way does it *actually* move? Explain.

activity 19
A full moon is just rising on the horizon.
a. Where will you see it in 1 hour? Why?
b. Where will you see it in 24 hours? Why?

activity 20 A
A lunar calendar has months that alternate between 29 days and 30 days so that each new month begins with a new moon. What is the disadvantage of using this calendar?

activity 20 B
How far do you move the Moon Marker on your lunar calendar to track the passing of 1 month? The passing of 1 year?

I

Answers (continued)

activity 6-7 B
Students should measure the length of each triangle in cm, then show that corresponding ratios are equal:
A/B = 3/4 = 0.75; a/b = 1.5/2 = 0.75.
c/b = 2.5/2 = 1.25; C/B = 5/4 = 1.25.

activity 8-9 B
a. 108 earths   b. 3.5 moons
c. 108 earths x 3.5 moons/earth = 378 moons

activity 8-10
The moon and sun have the same *apparent* size when seen from earth (about as big as a paper punch held at arm's length). But the sun is actually vastly larger than the moon. It appears equal because it is much farther away.

activity 8-11
a. diameter of earth =
   2 mm x 3.5 = 7 mm.
b. diameter of sun =
   7 mm x 108 = 756 mm = 75.6 cm.
c. distance between =
   75.6 cm x 108 = 8165.8 cm ≈ 8.2 m.

activity 9
The basketball has the same apparent size as the earth seen from the moon.

activity 9-10 A
The sun would appear as large as a paper punch at arm's length; the earth as large as a nickel at arm's length.

activity 9-10 B
1 cm/sun diameter x 108 sun diameters = 108 cm.

activity 10
Construct a pinhole projector: cover one end of a tube with foil and poke a pinhole in it. Cover the other end with waxed paper. Aim the pinhole at the sun so its light projects an image of the sun onto the waxed paper.

activity 10-11
View it from a distance of 108 basketball diameters.

activity 12 A
a. Eclipse the sun by covering it with your umbrella "moon."
b. No. This "moon" is proportionally larger than your head "earth."

activity 12 B
a. Yes. Half the moon is always illuminated (except during a lunar eclipse).
b. Half the crescent moon also remains in sun at all times. But from our viewing perspective we see only a thin crescent portion of the lighted side. The rest of the lighted half is "behind" the moon, hidden from our view.

activity 13 A
a. A little north or south of west.
b. A little north or south of east.
c. To the south or the north.
d. To the west.
e. To the east.

activity 13 B
New moon, waxing crescent, first quarter, waxing gibbous, full moon, waning gibbous, third quarter, waning crescent.

activity 14 A
Circles of latitude all lie in the same plane, ranging from largest at the equator to smallest around earth's poles. Circles of longitude, by contrast, all cross both poles, encompassing the earth's full circumference.

activity 14 B
a. The prime meridian.
b. The equator.
c. The ecliptic circle.

activity 14 C
equator, tropic of cancer, arctic circle

activity 15 A
Lean the north pole of the globe due north (or the south pole of the globe due south if you are in the southern hemisphere). Rotate it until the place where you live is on top.

activity 15 B
a. On the horizon.
b. Yes, during 2 brief periods every 24 hour hours, as the earth passes from day to night and night to day. (This may not be true for polar observers.)

activity 15-16
The sun appears to move from east to west across the sky. This apparent motion is caused by the earth's counterclockwise rotation, spinning the observer (and everything else) from west to east beneath the stationary sun.

activity 16-17
We experience changing seasons on earth because its axis tilts in different directions as it revolves around the sun. In northern winter, for example, the north pole tilts away from the sun, causing short days and long nights with the sun low in the south. After a quarter revolution around the sun, this axis is perpendicular to the sun's rays, creating equal days and equal nights with the sun culminating somewhat higher.

activity 17 A
a. The Tropic of Capricorn is a circle of latitude located at 23.5° S. It marks the southern-most boundary of the overhead sun. Further south, the sun always culminates to the north of your zenith.
b. The Arctic Circle is a circle of latitude located at 66.5° N. It marks the northern-most boundary of the setting and rising sun. Further north, observers experience periods of night or day that last longer than 24 hours.

activity 17 B
Winter solstice: N.P. tilts away from sun.
Spring equinox: N.P. tilts perpendicular to sun rays.
Summer solstice: N.P. tilts toward sun.
Fall equinox: N.P. tilts perpendicular again.

activity 18 A
It really rotates about 361°. The earth rotates full circle, then about 1° extra to bring the sun back to its 12 noon culminating position. This extra degree is needed to "catch up" with the sun that has drifted 1° east due to earth's revolution around the sun.

activity 18 B
No. In 24 hours of star time the earth turns 360° on its axis. In 24 hours of solar time the earth turns 361°. Thus, 24 solar hours are slightly longer than 24 star hours.

activity 3, 19
The moon appears to move from east to west due to earth's counterclockwise rotation on its axis. It actually moves in the opposite direction as it revolves counterclockwise around the earth. This eastward motion becomes apparent when viewing it over successive nights.

activity 19
a. In 1 hour the full moon will appear a little higher in the eastern sky due to the west-to-east rotation of the earth.
b. In 24 hours the moon will not have risen yet, due to its eastward revolution around the earth during that time period.

activity 20 A
The disadvantage is that 12 lunar months fall 11-12 days short of 1 solar year. Staying in step with the moon throws us out of step with the sun.

activity 20 B
Track the passing of 1 month by moving the Moon Marker 1 full circle around the earth. Track the passing of 1 year by moving the marker 1 full circle around the sun.

ACTIVITIES
AND
LESSON NOTES
1-20

☞ As you distribute these duplicated worksheets, **please observe our copyright notice** at the front of this module. We allow you, the purchaser, to make as many copies as you need, but forbid supplying photocopied materials to other teachers for use in other classrooms.

☞ TOPS is a small, not-for-profit educational corporation, dedicated to making great science accessible to students everywhere. Our only income is from the sale of these inexpensive modules. If you would like to help spread the word that TOPS *is* tops, please request multiple copies of our free **TOPS Ideas** catalog to pass on to other educators or student teachers. These offer a variety of sample lessons, plus an order form for your colleagues to purchase their own TOPS modules. Thanks!

AZIMUTH

1. Tie thread to a steel pin. Trim one end near the knot, cut the other end about 1 cm long. Slide the thread so the pin hangs level.

1 CM

2. Magnetize your pin: touch its *head* to the *south* pole of a magnet; its *point* to the *north* pole.

Then remove the magnet from your work area.

3. "Float" your pin *on* water in a small jar.

a. Slowly turn the jar. Does the pin turn with the jar? Explain.

b. Get a <u>Compass Circle</u>. Cut around the outside.
c. Ask your teacher where magnetized pinheads always point in your part of the world. Mark this *bearing* with a bold black dot just outside the center circle.

4. Cut a length of plastic straw as tall as your jar. Stick a lump of clay on one end.

a. Sink your floating pin. Use your "wand" to get it out again.

b. Rescue your pin this way whenever it sinks.

5. Cut an index card into a square, using a second one as a guide. Draw both diagonals on the square.

CUT ALONG EDGE

SQUARE

DRAW DIAGONAL LINES

6. *Center* your compass circle on the square with **N**, **S**, **E**, and **W** over the diagonals.

a. Tape it in this position.

b. Stick a pin through both layers, in the center. Enlarge the hole with a sharp pencil, then cut out the inside circle.

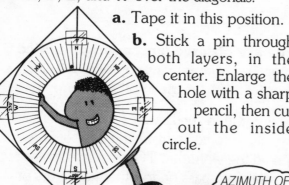

AZIMUTH OF CORNER A IS 165°.

7. Stick a raised "collar" of masking tape around the "shoulder" of the jar so half of its sticky side stays exposed.

a. Cut the raised tape into a fringe. Bend it outward all around the jar.

b. Set your compass circle onto this sticky fringe so it rests level.

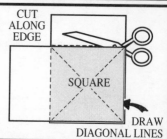

TAPE STICKS UP

CUT FRINGE

FAN OUT

8. "Float" your magnetized pin on the water. To find your bearings, always turn your compass circle so the pinhead points to the dot.

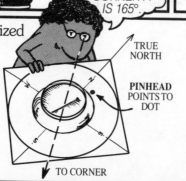

TRUE NORTH

PINHEAD POINTS TO DOT

TO CORNER

a. Find the *azimuth* (compass bearing) of corners A, B, C and D in your room from where you sit.

b. Clearly print these results on a small piece of paper. Fold it twice and toss it into a class "hat." *Don't* write your name on it.

TRY SEVERAL!

c. Draw someone else's paper from the hat. Use your compass to find whose desk it belongs to.

Objective

To build a compass. To define the location of your desk by finding the azimuth of each room corner.

Lesson Notes

Notice that steps 3a and 8a in this worksheet are <u>underlined</u>. Point out to your class that all underlined steps in each worksheet signal that a question should be answered, or that data should be entered, on a separate assignment sheet or in a personal science notebook.

2. Touching the pin as illustrated magnetizes the head of the pin *north* and its point *south*. It will keep this orientation unless it is accidentally remagnetized. Storing the permanent magnet out of reach avoids this potential problem, and also keeps its local magnetic field from influencing the compass pin.

3. The pin rests on top of the water, supported by surface tension. It doesn't truly float. Push it under, and the water displaced does not buoy it up again.

3b. Since the earth's *magnetic poles* do not coincide with its *geographic poles*, compasses show varying degrees of *declination*, as shown by the shaded angles below.

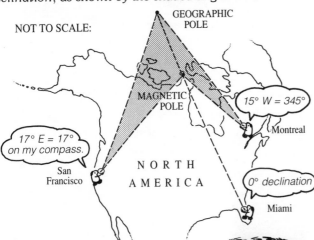

NOT TO SCALE:

GEOGRAPHIC POLE

MAGNETIC POLE

NORTH AMERICA

17° E = 17° on my compass.

San Francisco

15° W = 345°

Montreal

0° declination

Miami

Find your nearest city in the list below to estimate where magnetized pinheads point in your area. (All magnetic declinations have been translated into simple compass bearings.) Write your local bearing next to the head of a pin sketched on your blackboard. If you live outside of North America, ask a reference librarian.

354°

8. As students complete this step, check compasses to make sure all pinheads point in the same northerly direction. Any pin that points opposite was touched to the wrong pole in step 2, and should be remagnetized.

Metal desks may attract magnetized pins and skew compass readings. If this is a problem, tape the jars on top of inverted drinking cups to distance them from this influence.

Water in the jar should be changed daily to prevent surface stagnation that will eventually immobilize the magnetized pin.

Although students are not specifically directed to do so, they should identify this compass (plus all other instruments and models made in this TOPS module) with their names. These will be reused in activities that follow.

Answers

3a. No matter which way you turn the jar, the pin always points in the same direction.

8a. All azimuths should be recorded in degrees (°).

Materials

☐ Steel pins. Aluminum or brass pins *cannot* be magnetized.
☐ Thread and scissors.
☐ Any strong magnet. A "refrigerator" magnet is suitable. Make sure its poles are correctly labeled. (Hang it from a thread: the north pole will face north; the south pole, south.)
☐ The Compass Circle cutout, one per student. Students can work together in cooperative lab groups, but each one should make his or her own compass. These and other instruments will be used later for observations at home. Find this and all other reproducible cutouts in the back pages of this module, identified by activity number.
☐ A small baby food jar or equivalent, with a diameter that is small enough to accept the Compass Circle. If possible, select jars with well-defined "shoulders." Jars with a gradual slope are more difficult to fit with the tape fringe in step 7.
☐ Water.
☐ Your area's angle of declination marked on the blackboard.
☐ A straw and a small lump of clay.
☐ Two 4 x 6 inch index cards.
☐ Clear tape and masking tape.
☐ A pencil sharpener.
☐ Four room corners labeled A, B, C and D. Boldly write these letters on scrap paper or paper plates. Hang them where each corner meets the ceiling, so they can be seen from all desks. Leave these in place for the next 2 activities.
☐ A class "hat." If you're working with just one or two students, hide a penny for them to find. List the azimuths of each corner on scratch paper in advance.

Albany NY 347°	Cincinnati OH 0°	Honolulu HI 11°	Miami FL 0°	Pierre SD 11°	Santa Fe NM 13°
Amarillo TX 11°	Cleveland OH 356°	Hoquiam WA 22°	Milwaukee WI 1°	Pittsburgh PA 355°	Slt St Marie MI 356°
Anchorage AK 26°	Columbia SC 358°	Hot Springs AR 7°	Minneapolis MN 6°	Port Arthur ON 1°	Savannah GA 359°
Atlanta GA 1°	Columbus OH 358°	Idaho Falls ID 17°	Mobile AL 4°	Portland ME 343°	Scranton PA 350°
Atlantic Cty NJ 350°	Dallas TX 8°	Indianapolis IN 1°	Montgomery AL 3°	Portland OR 21°	Seattle WA 22°
Austin NV 17°	Denver CO 13°	Jackson MI 5°	Montpelier VT 345°	Providence RI 345°	Shreveport LA 7°
Baker OR 20°	Des Moines IA 7°	Jacksonville FL 0°	Montreal QE 345°	Quebec QE 341°	Sioux Falls SD 9°
Baltimore MD 352°	Detroit MI 357°	Juneau AK 29°	Moose Jaw SK 17°	Raleigh NC 356°	Sitka AK 28°
Bangor ME 341°	Dubuque IA 4°	Kansas City MO 8°	Nashville TN 2°	Reno NV 18°	Spokane WA 22°
Birmingham AL 3°	Duluth MN 5°	Key West FL 1°	Needles CA 15°	Richfield UT 16°	Spgfld IL 4°
Bismarck ND 12°	Eastport ME 339°	Kingston ON 348°	Nelson BC 22°	Richmond VA 354°	Spgfld MA 346°
Boise ID 19°	El Centro CA 14°	Klmth Flls OR 19°	New Hvn CT 347°	Roanoke VA 356°	Spgfld MO 7°
Boston MA 345°	El Paso TX 12°	Knoxville TN 359°	New Orleans LA 6°	Sacramento CA 17°	Syracuse NY 349°
Buffalo NY 352°	Eugene OR 20°	Las Vegas NV 15°	New York NY 349°	St John NB 338°	Tampa FL 1°
Calgary AB 22°	Fargo ND 9°	Lewiston ID 20°	Nogales AZ 13°	St Louis MO 5°	Toronto ON 353°
Carlsbad NM 12°	Flagstaff AZ 14°	Lincoln NB 9°	Nome AK 17°	Salmon ID 19°e	Trinidad CO 13°
Charleston SC 358°	Fresno CA 16°	London ON 355°	N Platte NE 11°	Salt Lake City UT 16°	Victoria BC 23°
Charleston WV 357°	Garden City KS 11°	LA CA 15°	OK Cty OK 9°	San Antonio TX 9°	Watertown NY 348°
Charlotte NC 358°	Grand Jnctn CO 15°	Louisville KY 1°	Ottawa ON 347°	San Diego CA 14°	Wichita KS 9°
Cheyenne WY 13°	Grnd Rpds MI 359°	Mnchstr NH 345°	Phila PA 350°	San Francisco CA 17°	Wilmington NC 356°
Chicago IL 1°	Helena MT 19°	Memphis TN 5°	Phoenix AZ 14°	San Juan PR 352°	Winnipeg MB 9°

ALTITUDE

1. Get a <u>Quadrant</u> sheet. Fold this whole paper across the middle, precisely along the dotted line.

2. Keeping the paper folded, carefully cut around the remaining 3 sides of the rectangle.

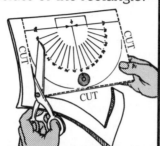

3. Tie thread to a washer. Trim one end short.

a. Hold the thread and washer directly over the 0° line.

b. Tape the thread as shown where the arrows meet. Trim excess thread.

TAPE NEATLY.

WASHER ON THREAD

4. Fold both layers precisely along the 90° line and make a sharp crease. Keep the word "quadrant" face up.

CREASED FOLD

QUADRANT

a. Anchor a long piece of *sticky-side-up* tape along all four *edges* as shown.

CREASE

TAPE

b. Center a straw over the sticky tape so it lines up with the guides at each end.

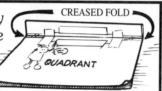

c. Press the straw firmly against the spine of a book, then stick both flaps together with rolled tape.

QUADRANT

JOIN FLAPS WITH ROLLED TAPE

5. Get the <u>Angles of Altitude</u> cutout. Fold it along the dotted line, then hang it over a level table, anchored with a book.

a. Use your *quadrant* to measure each angle of *altitude*: to the treetop, top of tower, jet plane, cloud, and bird.

. . . and the tower's altitude is . . .

b. What is the altitude of your *horizon*? Your *zenith*?

c. Cut out the <u>Protractor</u> and measure each angle of altitude again. What do you find?

d. When astronomers measure the altitude of an object, do they measure its height? Explain.

6. Pick *one* of your room corners: A, B, C *or* D. Aim your quadrant at this corner from where you sit.

Ⓑ

SIGHT THROUGH THE STRAW.

a. Make 10 careful measurements of this *same* corner. Write them in a row.

b. You can't aim your quadrant and read its altitude at the same time. How did you solve this problem?

c. Can you ever find the *exact* altitude of the corner from where you sit? Explain.

d. Find the altitude of your shoes. Why should this reading have a minus sign?

Objective

To construct a quadrant. To practice measuring the altitude of a room corner from where you sit.

Supporting Concepts

We have worked hard to make our worksheet directions easy to follow, within the grasp of any student with basic reading skills. But reading comprehension, by itself, is not enough. Your students also need to understand *supporting concepts* to help them interpret and explain what they do and see. Depending on the needs and abilities of your particular age group, you might decide to treat every concept, cover a select few, or skip this section entirely. Try to encourage maximum learning independence with a minimum of teacher intervention.

➊ Think of a straw as a mini-telescope. Practice looking through it to a wall clock or other target object in your room. Be careful not to poke the straw in your eye.

• Notice that your "telescope" is easier to aim if you keep one eye closed.

• Notice that your eyes feel more rested if you open both after finding your target.

➋ Make a demonstration quadrant in advance by completing steps 1-4. Decide with your class the best way to hold the quadrant while aiming the straw at a target object.

• If you tilt the face of the quadrant *away* from the washer, the washer takes too much time to stop swinging.

• If you tilt the face of the quadrant *into* the washer, it can't swing freely.

• If you hold the face of the quadrant *almost vertical* so the washer lightly brushes the paper, it swings freely *and* quickly comes to rest. This is the proper orientation. (Do not discuss how to read the quadrant at this time. Students will consider this in step 6b.)

Lesson Notes

1-4. Stress the importance of doing a good job. Careful folding, cutting and taping produces a quadrant that measures true.

4c. Pressing the straw against the book spine before taping both flaps of the quadrant together prevents gaps from forming around the fold lines.

Answers

5a. Expect measuring error deviations of ±1°: Altitude to the…

treetop = 10°
top of the tower = 25°
jet plane = 52°
cloud = 77°
bird = 90°

5b. (The horizon and zenith are both defined on the Angles of Altitude cutout.) Altitude of your…

horizon = 0°
zenith = 90°

5c. All angles have the same value when measured by the protractor or by the quadrant. Both instruments measure in the same units of degree.

5d. No. In astronomy, the altitude of an object refers to the angle it makes with your horizon, not to its height above the ground.

6a. Altitudes vary with desk location. Expect measuring error to diminish with practice. For example: corner C = 25°, 32°, 30°, 27°, 23°, 26°, 29°, 28°, 26°, 27°.

6b. Here are three possible solutions:

Buddy system: One person aims the quadrant. After the washer settles, the other reads the altitude on its face. (This method is both easy and accurate, making it especially useful for younger children with less eye-hand coordination.)

Aim-and-tip method: Aim the quadrant. After the washer settles, tip the quadrant's face into the washer to hold it in place while slowly turning to read the altitude. (This method allows solo observations, but requires sophisticated eye-hand coordination, especially at steeper altitudes.)

Aim-and-hold method: Aim the quadrant while *almost* holding the washer between thumb and forefinger. After the washer settles, pinch it in place and read the altitude. (This method works equally well at any altitude. We like it best.)

6c. No. With practice and good measuring technique you can limit measuring error to a small range (perhaps ± 1°) but never eliminate it entirely.

6d. Varied answers. The altitude is negative because your shoes normally remain on the floor, below your horizon.

Students should write their names on their quadrants and save them for future use. The Angles of Altitude cutout and Protractor cutout may be discarded. Better yet, save them for the next time you teach this module.

Materials

☐ The Quadrant sheet, one per student.
☐ Scissors.
☐ Thread.
☐ A medium-sized washer. The size illustrated on the quadrant cutout has a 3/4 inch (19 mm) outside diameter.
☐ Clear tape.
☐ Clear tape that is sticky on both sides (optional). If you choose to use this, your students won't need to anchor the tape around its perimeter in step 4a, or roll it sticky-side-out in step 4c.
☐ A straw. Don't use large diameter milk shake straws or flexible straws.
☐ A heavy book.
☐ The Angles of Altitude cutout.
☐ The Protractor cutout.

LUNAR COORDINATES

1. From where you sit, find the *coordinates* of the 4 corners in your room labeled A, B, C and D:

THESE ARE COORDINATES. THEY TELL WHERE A POINT IS LOCATED.

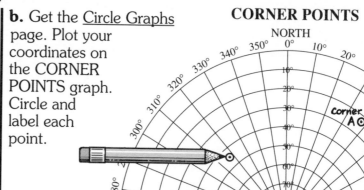

Measure the **azimuth** with your *compass*. Write it first.

(xx°, yy°)

Measure the **altitude** with your *quadrant*. It goes second.

a. List your data in a table.

COORDINATES

CORNER	(Azimuth, Altitude)
A	
B	
C	
D	

b. Get the <u>Circle Graphs</u> page. Plot your coordinates on the CORNER POINTS graph. Circle and label each point.

CORNER POINTS

NORTH

Corner A⊙

c. Which corner has the greatest altitude? Does this mean it is higher than the other corners? Explain.

---

2. Take home your compass and quadrant to track the moon's changing coordinates across the sky. List your data in tables like these:

SIGHT COMPASS TO HORIZON POINT.

Take FLASHLIGHT, COMPASS & QUADRANT.

a. Begin your ONE WEEK observations *each night* at the *same hour*. Start when the moon is in the WESTERN (sunset) half of the sky. Record at least 3 moon positions, many more if possible.

b. Begin your ONE NIGHT observations *at sunset*. Start when the moon is in the EASTERN (sunrise) half of the sky. Pick a clear night when you can make 4 or more observations in a row about an hour apart.

ONE WEEK

TIME (1 only)	DATES (3 minimum)	COORDINATES (Azimuth, Altitude)

ONE NIGHT

DATE (1 only)	TIMES (4 minimum)	COORDINATES (Azimuth, Altitude)

---

3. Find your <u>Circle Graphs</u> page.

a. Plot and circle each table of data on the correct graph. Label beginning and end points with the correct date or hour.

b. Sweep a *smooth* line among your data points.

c. Why are your data points scattered to some extent?

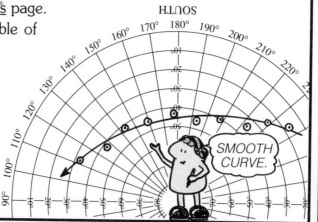

SMOOTH CURVE.

4. How did the moon appear to move across the sky…

a. over ONE WEEK?

b. over ONE NIGHT?

E to W? W to E?

Objective

To measure the azimuth and altitude of the moon as it appears to move across the night sky. To plot this data on circular graphs.

Supporting Concepts

❶ If your desks are arranged in rows, identify desk positions as an ordered pair of coordinates. First define the coordinate system (from your students' point of view) on your blackboard. Then ask volunteers to write their own unique desk positions underneath:

(row , position)

number left to right, beginning with 1.

number front to back, beginning with 1.

End with a chain activity: the teacher calls out a desk coordinate; the student at that coordinate position calls out another desk coordinate; the student at that new position calls out another, and so on.

❷ Demonstrate graphing technique on your blackboard: Connect various sets of circled data points with the best possible smooth line. Notice that…

• The data points are scattered, yet the line is smooth, seeking an "average" trend.

• The line never passes through a circle, so as not to obscure its data point inside.

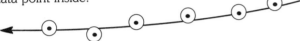

Lesson Notes

1. This step allows your students to practice the measuring skills they'll need to gather moon data at home (step 2) and the graphing skills they'll need to complete step 3. Depending on the day in your calendar month, you may wish to assign step 2 immediately as homework, or wait for the start of the next moon cycle. (Read on to decide.) In either case, students may start activity 4 at any time.

2. Prepare your class for a successful homework experience by reviewing these basic requirements.

WHAT is needed:

• a compass, a quadrant and a flashlight.

• two observation tables. (Students can draw these now in class, or later at home, using this worksheet as a reference.)

WHEN to begin:

• Start the ONE WEEK table within 2-10 days of the new moon, the sooner the better. Start after 10 days only if you are a night owl. Otherwise, wait for the next moon cycle.

• Start the ONE NIGHT table at sunset on any clear night within 8-15 days after the new moon. Past 15 days, the moon rises past sunset, pushing the string of 4 or more observations late into the evening.

If bad weather doesn't allow minimum observation opportunities for either table, wait until the next moon cycle to complete one or both tables again. Your students won't use the results of this activity until they reach lesson 19.

HOW to take accurate measurements:

• Line up your compass on the ground. Leave it in place for multiple moon observations. (Even if the pin sinks, the compass remains aligned.)

• If the moon is high overhead, first sight straight down to a point near your horizon. Read the azimuth of that horizon point to estimate the azimuth of the moon. Confirm your reading is correct by sweeping your gaze straight up, from

azimuth back to moon. (This technique allows ±5° accuracy.)

• Use the aim-and-hold method to measure altitude. Repeat quadrant readings until they agree. (This technique allows ±1° accuracy.)

Answers

1. Data tables and graphs will vary, depending on each student's unique desk position. The corner (or corners) with greatest altitude are not higher off the floor, but closer to the observer. (A student at the center of your room, equidistant from all 4 corners, will graph them at equal altitudes.)

2-3. Your results will differ from these sample graphs, depending on your latitude, season of year, time of night, and position in the moon cycle. An azimuth uncertainty of ±5°, and an altitude uncertainty of ±1° will scatter the data points.

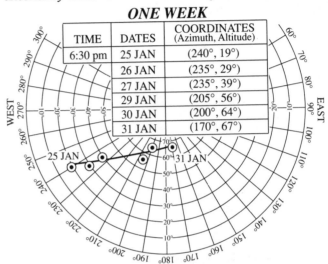

ONE WEEK

TIME	DATES	COORDINATES (Azimuth, Altitude)
6:30 pm	25 JAN	(240°, 19°)
	26 JAN	(235°, 29°)
	27 JAN	(235°, 39°)
	29 JAN	(205°, 56°)
	30 JAN	(200°, 64°)
	31 JAN	(170°, 67°)

ONE NIGHT

DATES	TIME	COORDINATES (Azimuth, Altitude)
30 JAN	6:30 pm	(200°, 64°)
	7:30 pm	(225°, 62°)
	8:30 pm	(242°, 50°)
	9:30 pm	(257°, 40°)
	10:30 pm	(272°, 31°)

3c. The data points are scattered because of unavoidable measuring errors in aligning the compass and quadrant.

4a. Over one week, the moon moves from west to east, rising to its highest altitude in the south.

4b. Over one night, the moon appears to move from east to west, again rising to its highest altitude in the south.

Materials

☐ Corner positions labeled A, B, C, and D from activity 1.
☐ The compass and quadrant previously constructed.
☐ The Circle Graphs sheet.

SOLAR SURVEY

1. Take your compass and quadrant outside. Practice these *indirect* ways of measuring the sun's position *without looking directly at it!*

a. AZIMUTH: Align the pinhead and dot with your jar level. Invert your quadrant so its thread casts a shadow. This marks the sun's azimuth when aligned with the radial lines on your compass.

b. ALTITUDE: Hold the straw so sunlight shines through it onto your hand, forming a *small circle.* Keep practicing until you get consistent altitude readings.

READ AZIMUTH HERE.

A HANDY PLUMB LINE!

SET JAR LEVEL

SHADOW ACROSS CENTER

SUN RAYS

THE SHADOW'S ALMOST GONE . .

CIRCLE OF LIGHT

Caution: protect your precious eyesight! **Never** look directly at the sun.

---

2. List your sun coordinates, and the time you measured them, in a table.

a. Continue tracking the sun's apparent motion across the sky all day long. Try to space your observations about an hour apart.

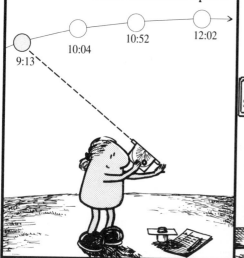

9:13 10:04 10:52 12:02

TIME	SUN COORDINATES (Azimuth, Altitude)

b. If time gaps develop in your data record, fill them in promptly on the next sunny day. Try to capture a sunset or sunrise (or both) at home.

BUS STOP

THE THINGS I'LL DO FOR SCIENCE . . .

3. Cut the <u>Circle Graph</u> page for this activity in half. Use the part titled "Solar Survey."

a. Plot and circle each set of sun coordinates.

b. Draw a smooth, sweeping line among the circles. Label the times of your earliest and latest sightings.

4. Study the graph line you have just drawn. Answer these questions:

<u>a.</u> Where did the sun rise and set?

<u>b.</u> Where did it *culminate* (reach its highest point)? Did it pass straight overhead?

<u>c.</u> Summarize how the sun appears to move across the sky.

Objective

To indirectly measure the azimuth and altitude of the sun by casting shadows. To plot this data on a circular graph.

Supporting Concepts

✪ Because the sun is so far away, its light rays are moving essentially parallel when they strike our earth. At any given instant, observers on your school grounds (or in your town) see sun shadows pointing in the same direction.

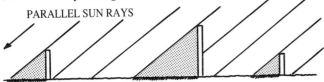

PARALLEL SUN RAYS

Lesson Notes

You can't track the sun on a cloudy day. If overcast conditions prevail, proceed directly to activity 6. Return to activities 4 and 5 when the weather clears.

1. This step allows students to practice measuring the azimuth and altitude of the sun by observing shadows. Self-directed students might do this independently. Or you can complete this step as a teacher-directed activity. Either way, mastery of both measuring techniques is essential to success in the steps that follow.

1a. When properly aligned, the thread's shadow crosses the *center* of the jar.

2. Your students are now ready to enter their local time and sun position in the top line of a data table. Completing the rest of this table is no small task. Your students must remember to make periodic sightings throughout the day as schedule and weather allow, then continue collecting data at home. Sunny weekends provide the best opportunity to record coordinate positions over a wide time range.

Answers

2-3. Sample Result: 45° N latitude, late January:

TIME	COORDINATES (Azimuth, Altitude)
12:25 pm	(191°, 24°)
1:25 pm	(207°, 22°)
2:25 pm	(212°, 16°)
3:05 pm	(225°, 13°)
3:40 pm	(227°, 8°)
4:45 pm	(240°, 0°)
8:10 am	(130°, 7°)
9:15 am	(144°, 16°)
10:25 am	(163°, 22°)
11:55 am	(180°, 26°)

Because of differences in latitude and time of year, this sample result will not likely match the experimental results of your students, except for the scattering of data points.

To anticipate the shape and placement of your local sun track, locate the line corresponding to your current month in the graph below. For every degree of latitude that you live further north, *drop* the entire graph line 1° to the southern horizon. For every degree of latitude that you live further south, raise the entire graph line 1° toward the zenith.

Theoretical Sun Tracks: 42° N latitude, about the 21st of each month:

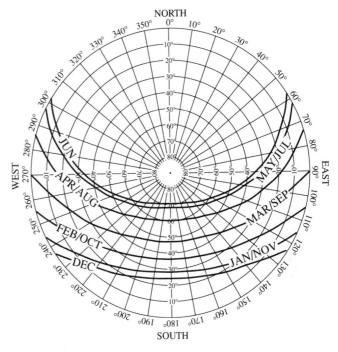

4a. If your students captured a sunrise or sunset at home, they can reports its azimuth with accuracy. Without direct data they can still roughly estimate these azimuth points by extrapolating (extending) their graph line to 0° latitude at both ends. (In summer or winter your class may be surprised to find the sun rising and setting considerably to the north or south of due east and west. This variation, always the same at both horizons, will be examined in activity 17.)

4b. The sun culminates due south at an azimuth of 180°. In the US and Canada, the sun never passes straight overhead, though it comes close during summer solstice in southern Florida and Texas.

4c. The sun appears to rise in the east (not necessarily due east), culminate in the south and set in the west (not necessarily due west).

Extension

Graph the azimuth of the rising or setting sun against time as measured in weeks. (Students who complete this year-long science project will become astute observes of the seasons.)

Materials

☐ A sunny day.
☐ Your compass.
☐ Your quadrant.
☐ A watch.
☐ The Circle Graph labeled "Solar Survey."

SHADOW TRACK

1. Cut out the <u>Sundial Circle</u>. Center it on the *back* of a paper plate, held with *small* rolls of masking tape.

ROLLED TAPE

2. Poke a pin-hole through the center mark.

Repoke the pin up from below, and tape the head in place.

USE THE SAME HOLE.

TAPE OVER PINHEAD

3. Nearly fill a can with dry gravel or sand.

a. Stick masking tape around the top rim so half of the sticky side remains free. Cut the exposed tape into fringes, to the can's rim.

b. Fan these outward, then center the paper plate on top to make a *sundial*.

CUT TAPE FRINGE

FAN OUT

GRAVEL

4. Hold the pin straight up with a *small* lump of clay. This clay should not extend outside the smallest 80° circle.

5. Find a place that receives direct sunlight most of the day. Take your sundial, compass and quadrant with you.

a. Level the *sundial* with the quadrant in 2 perpendicular directions. Adjust the pin, as necessary, to point straight up.

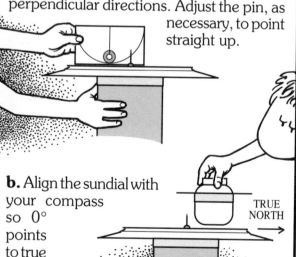

b. Align the sundial with your compass so 0° points to true north.

TRUE NORTH

c. Try to leave your sundial in place, without disturbing it in any way, for the rest of the day.

6. You can use the tip of the pin's shadow to find *both* the sun's azimuth and altitude!

a. Does the shadow point to, or directly opposite, the azimuth? How can you calculate it?

b. Does the tip of the shadow show the altitude? Draw a sketch (side view) to support your answer.

c. Use your sundial to track the sun's journey throughout the day (on an hourly basis, if possible). Organize your data in a table, as before.

7. Graph your data on the <u>Circle Graph</u> labeled "Shadow Track."

a. Compare this sun graph with your "Solar Survey" graph. How are they similar? How are they different?

RRRINGGG!

b. How could you make a sundial that tells time?

Objective

To measure the azimuth and altitude of the sun with a sundial. To plot this data on a circular graph.

Supporting Concepts

❂ Stick the point of a pencil in a lump of clay. Stand it vertically on the floor in front of your students. Turn out the lights, then shine a flashlight on it from above.

Think of yourself as a tiny observer standing near the lump of clay. Discuss these questions with your class:

• How is the azimuth of the pencil shadow related to the azimuth of the flashlight? (The shadow always lies 180° opposite the flashlight.)

• How is the length of the pencil shadow related to the altitude of the flashlight? (The shorter the shadow, the greater the altitude of the flashlight.)

Lesson Notes

1. After centering the sundial circle over the plate, there is a natural tendency to set the tape by pressing it flat against the table. To avoid caving in the paper plate, your students should first turn it flat-side-down. A warped sundial will warp their results.

5-7. Your students should locate their sundial in a low traffic area with good solar exposure to the east, south and west. Once properly sighted, leveled and aligned, the azimuth and altitude of the sun can be read at the tip of the pin's shadow without touching the sundial again.

If conditions on your school grounds are not so ideal, consider these alternatives:

(a) Ask students to complete this activity at home over a sunny weekend.

(b) Set up just one sundial for your entire class to read (but not touch). Individuals, lab teams or the entire class can record data and generate graphs.

(c) Move the sundial as necessary to keep it in continuous sunlight. Here's how:

Before it is overtaken by shade, accurately mark where the tip of the pin's shadow "touches" the sundial circle.

Quickly move the sundial to its new sunny location. Rotate it so the tip of the pin again casts its shadow to the mark you just made. Use your quadrant to make sure it again rests perfectly level.

Answers

6a. The sun and its pin shadow have azimuths that are direct opposites. Add *or* subtract 180° from the shadow azimuth to compute a sun azimuth that lies between 0° and 360°. For example:

If the shadow points to 350°, the sun azimuth is:
350° − 180° = 170°.
You can't add, because the result would be greater than 360°.

If the shadow points to 10°, the sun azimuth is:
10° + 180° = 190°.
You can't subtract, because the result would be less than 0°.

6b. Yes. The shadow's tip points to altitude circles between 20° and 80°.

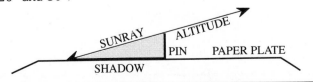

6c. The sun's apparent position in the sky varies with latitude and the time of year. This sample result was taken at 45° N latitude in late January.

7.

TIME	SHADOW'S AZIMUTH	SHADOW'S ALTITUDE	COORDINATES (Azimuth, Altitude)
10:50 am	349°	22°	(169°, 22°)
11:50 am	4°	24°	(184°, 24°)
12:45 pm	16°	25°	(196°, 25°)
1:45 pm	32°	23°	(212°, 23°)
2:20 pm	40°	20°	(220°, 20°)

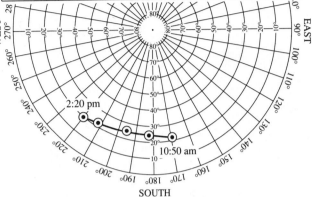

7a. Similar: Both graphs lines have the same general shape and placement. (There is very little change in the apparent path of the sun across the sky from one day to the next.) Different: The data points generated by reading this fixed sundial don't scatter as widely across the graph, suggesting less measuring error.

The sun track no longer culminates due south, as it should. Error in the original placement of the sundial (step 5) now repeats consistently with *every* reading. (Measuring error that was random in the last experiment has now become locked in, or *inherent,* in this experiment.)

Morning and evening sun altitudes can no longer be read because the pin shadow lengthens beyond its limiting 20° altitude ring.

7b. Get a fresh paper plate. Calibrate the face as a sundial by marking and labeling the shadow's point position at regular intervals (hours, half hours, etc.). On subsequent sunny days, read your local time at the tip of the shadow.

Extension

Calibrate a sundial in hours. Does your instrument keep accurate time in all seasons?

Materials

☐ The Sundial Circle. See notes 5-7 for quantity.
☐ Scissors.
☐ A paper plate. See activity 20 for specifications.
☐ A 1 inch (2.54 cm) straight pin. Altitudes on the Sundial Circle are calibrated to this length, a standard size in U.S. industry. If you can't find this length, cut longer pins to size.
☐ A medium or large can.
☐ Dry gravel or sand.
☐ Masking tape.
☐ Modeling clay.
☐ A sunny day.
☐ Your compass.
☐ Your quadrant.
☐ A watch.
☐ The Circle Graph labeled "Shadow Track."

OFF THE WALL

1. Carefully cut out a <u>Measuring Triangle</u>. Tape it even with the corner of an index card like this:

KEEP TAPE OFF THE CARD EDGES.

Your Name

LINE UP EDGES

ROLLED TAPE

2. There are 3 squares of newspaper taped together on your wall: square **A** measures 30 cm on each side; square **B** measures 40 cm; square **C** measures 50 cm.

a. Accurately make a *scale drawing* of this figure on notebook paper. Make all *corresponding* sides 10 times shorter than actual size.

b. Label each square with corresponding lower case letters: *a*, *b* and *c*.

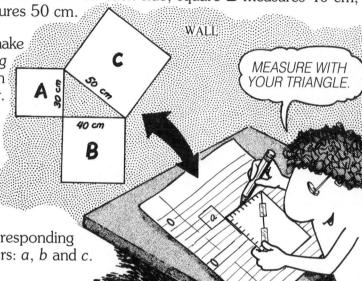

WALL

C

A 30 cm

50 cm

40 cm

B

MEASURE WITH YOUR TRIANGLE.

3. Rewrite each *ratio* below as a decimal. What can you discover about corresponding sides?

a. a/b, A/B
b. A/C, a/c
c. c/b, C/B
d. A/a, B/b, C/c

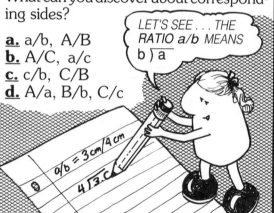

LET'S SEE . . . THE RATIO a/b MEANS b)‾a

a/b = 3cm/4cm

4)‾3.c

4. A drawing is *to scale* if corresponding parts have the same ratio.

a. Did you draw the newspapers on the wall *to scale*? How do you know?

b. Calculate the *diagonal* of the largest newspaper square without measuring it. Explain how you did this.

c. Check your answer.

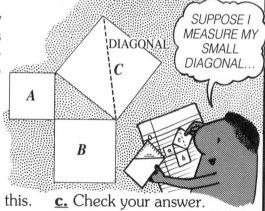

DIAGONAL

C

A

B

SUPPOSE I MEASURE MY SMALL DIAGONAL...

5. Make another scale drawing 100 times smaller than actual size.

a. Label the corresponding parts with lower case letters.

b. Can you draw a figure scaled down 1,000 times? Explain.

10 times smaller:

c

a

b

MARK THE SCALE OF EACH DRAWING.

100 times smaller:

6. Choose another scale that fills the whole back of your notebook paper. (Hint: think about enlarging an earlier drawing.)

a. Now draw it. How many times smaller is it than the original?

b. Use ratios to show that you have again drawn the newspapers on the wall *to scale*.

FILL YOUR PAGE.

c

a

b

Objective

To draw a geometric figure to scale. To observe that corresponding parts in the scale drawing have the same ratio as corresponding parts in the original.

Supporting Concepts

✪ Demonstrate how to accurately draw right angles using an index card with square corners. To draw a stairway, for example, always align the card with a previous line *before* drawing a new line.

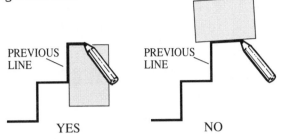

PREVIOUS LINE

PREVIOUS LINE

YES   NO

✪ To divide a number by 10, 100, or 1000, move its decimal 1, 2, or 3 places to the left:
53/10 = 5.3;   53/100 = 0.53;   53/1000 = 0.053.

✪ A ratio compares two things. It can be expressed a fraction: 1/2, 3/4, etc., are ratios.

✪ Convert ratios to decimals by dividing the lower half of the fraction (the denominator) into the top half (the numerator):
$$3/4 = 4\overline{|3} = .75$$

✪ A drawing is *to scale* if its corresponding parts have the same ratio as the original:

Draw this triangle on your blackboard. Use an index card as your unit of measure. Number each side with the correct length and show all unit marks. Label it "original".

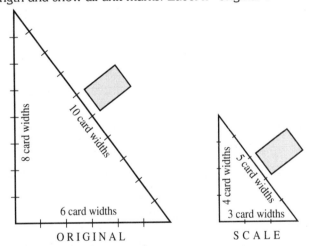

8 card widths
10 card widths
6 card widths
ORIGINAL

4 card widths
5 card widths
3 card widths
SCALE

Ask a volunteer to *scale* this original drawing 2 times smaller. Show that corresponding parts in both triangles have equal ratios.
3/4 = 6/8 = .75;   10/8 = 5/4 = 1.25;   etc.

Lesson Notes

2. Students should produce a scale drawing, not a freehand sketch. Insist that they measure accurately and draw right angles that are true.

4b. This step calls for a prediction supported by calculations. Complete it *before* measuring the diagonal of square C with a meter stick in the next step.

Answers

2. Students should draw and label 3 squares on their notebook paper to form a central right triangle that measures 3, 4, and 5 cm on a side.

3. Corresponding sides of the scale model and the newspapers on the wall have the same value:
   a. a/b = 3/4 = 0.75;   A/B = 30/40 = 0.75.
   b. A/C = 30/50 = 0.6;   a/c = 3/5 = 0.6.
   c. c/b = 5/4 = 1.25;   C/B = 50/40 = 1.25.
   d. A/a = 30/3 = 10; B/b = 40/4 = 10; C/c = 50/5 = 10.

4a. This drawing is properly scaled to the newspapers on the wall because corresponding sides have the same ratio.

4b. The diagonal of scaled square c measures 7.1 cm. Its corresponding diagonal on newspaper square C is 10 times longer, or 71 cm.

4c. Agreement should be close.

5, 5a.

a   c
b

5b. Encourage students to debate this question: the squares would measure 0.3 mm, 0.4 mm and 0.5 mm on a side. This is large enough to see, but not large enough to draw with any accuracy, given an instrument as blunt as a pencil.

6a. On the reverse side of their notebook paper, students should draw squares measuring 6, 8 and 10 cm on a side. These are 5 times smaller than the original newspaper squares on the wall. (Or 2 times *larger* than the squares in step 2.)

6b. Students should again show that corresponding ratios are equal. Here are two examples:
   a/b = 6/8 = 0.75;   A/B = 30/40 = 0.75.
   C/A = 50/30 = 1.67;   c/a = 10/6 = 1.67.

Materials

☐ The Measuring Triangle cutout.
☐ Scissors.
☐ A 4 x 6 inch index card or larger.
☐ Tape.
☐ The newspaper wall display detailed in step 2. Prepare this in advance. Label the squares as shown with lettering that is clearly visible from all parts of your room.
☐ A meter stick.

BIRD'S EYE VIEW

1. Think of the *main floor* area in your classroom as a large rectangle. Ignore "nooks and crannies" or attached rooms that don't fit your rectangle.

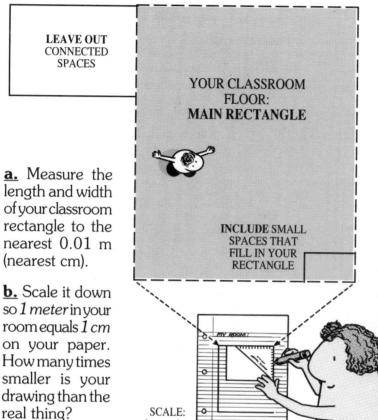

LEAVE OUT CONNECTED SPACES

YOUR CLASSROOM FLOOR: **MAIN RECTANGLE**

INCLUDE SMALL SPACES THAT FILL IN YOUR RECTANGLE

a. Measure the length and width of your classroom rectangle to the nearest 0.01 m (nearest cm).

b. Scale it down so *1 meter* in your room equals *1 cm* on your paper. How many times smaller is your drawing than the real thing?

MY ROOM:

SCALE: 1 cm = 1 m

c. Show that corresponding sides of your scale drawing and your room have equal ratios.

2. Imagine that you are looking down into your room from above.

a. Draw, *to scale*, some of the room's major features. Label each item.

BLACKBOARD

TEACHER'S DESK

MY ROOM
SCALE: 1 cm = 1 m

MY DESK

DOOR

b. A desk is located away from walls, somewhere near the center of your room. What measurements must you find to add it to your scale drawing?

3. On a scrap of paper, draw yourself to the same scale as your room. A stick figure is fine, as long as you accurately scale your height.

YOUR HEIGHT TO SCALE

4. Trim away *all* extra paper from three sides of your figure. Leave a "handle" on one side. Check that its width still equals your scale height.

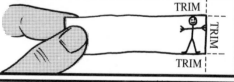

TRIM

TRIM

TRIM

5. Use this small figure like a ruler on your drawing to *predict* how many times the real you fits, head to heel, across your real classroom.

a. Test your prediction. How accurate was it?

b. Show that your scale body and scale room are *proportional* (have the same ratio) as your real body and real room.

PROPORTIONAL MEANS CORRESPONDING PARTS HAVE THE **SAME RATIO.**

ONE LENGTH   TWO LENGTHS

Objective

To draw your classroom, yourself, and other familiar objects to scale. To observe how corresponding parts in a scale drawing have the same proportion as the actual object.

Supporting Concepts

❖ A meter stick (1 meter long) contains 100 centimeters. To convert any measurement in centimeters to meters, therefore, divide by 100. (Move the decimal 2 places to the left):

100 cm = 1.00 m,
70 cm = 0.70 m,
25 cm = 0.25 m,
57 cm = 0.57 m.

Measure the length of your blackboard to the nearest cm. Express your answer in meters plus centimeters, then simplify. Suppose, for example, the meter stick fits 4 times with 27 cm left over:

blackboard length =
4 meters + 27 cm = 427 cm = 4.27 m

❖ Lines are *parallel* if they run side by side without crossing, no matter how long you extend them. Lines are *perpendicular* if they are at right angles to each other.

PARALLEL

PERPENDICULAR

❖ Objects are *proportional* if corresponding parts have the same ratio:

Which rectangles are proportional?

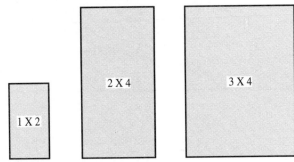

2 X 4

3 X 4

1 X 2

$1/2 = 2/4 \neq 3/4$. A and B are proportional, but not C.

Lesson Notes

2a. These room features should be *drawn to scale,* not simply sketched in. This requires a determination of both their *location* and *size.* Consider, for example, the problem of drawing a free-standing desk.

LOCATION: Measure the perpendicular distance from a desk corner to a point on a wall; then measure from this wall point to a room corner. On your drawing, make scale measurements in reverse order to establish the location of the desk corner within your scale drawing.

SIZE: Measure the width and length of the desk. Draw it parallel to the walls of the room from the corner point just established. If the desk is not at right angles to the walls, a second corner location may be needed.

While precise placement of objects is not mandatory, accuracy in measuring and drawing are important skills worth developing.

3-4. Students should measure their height to the nearest cm with a meter stick against a wall. If someone's height measures 1 meter, 46 cm (1.46 m) for example, the scale height of their stick figure is 1.46 cm after trimming.

5. Because the "body ruler" is drawn to scale, its unit length fits the same number of times across the scale drawing as the real body fits across the real classroom. To check this, one student should repeatedly stretch out on the floor (or stretch a string cut as long as body height) while another marks end-to-end body lengths. Remind students who choose to lie face down not to stretch out their toes. The body ruler is scaled to flat-footed height.

Answers

1. These answers are based on a sample classroom that is 12.08 m long and 8.53 m wide:

1a. length = 12.08 m, width = 8.53 m
1b. The scale drawing is 100 times smaller. (Students should draw a rectangle scaled to 12.08 cm by 8.53 cm.)
1c. Corresponding ratios are equal.
    drawing: l/w = 12.08 cm / 8.53 cm = 1.42
    room: l/w = 12.08 m / 8.53 m = 1.42

2a. At least some of the objects students choose to draw should be located away from the walls.

2b. Needed measurements:
• The width and length of the desk top.
• The perpendicular distance from a desk corner to the nearest wall.
• The distance from this wall point to the nearest corner of the room.

5-5a. Students will typically report some whole number of body lengths plus a fraction more. Agreement at this level of accuracy should be excellent. (If a real body fits significantly fewer times across the room than a scale body, ask if pointed toes may have added extra length.)

5b. Students should write ratios between corresponding parts and show that they are equal. Here are a few possibilities. Encourage diversity:

real body height/real room length =
                scale body height/scale room length
scale desk width/scale body height =
                real desk width/real body height
real door width/real room width =
                scale door width/scale room width

Materials

☐ A meter stick.
☐ The measuring triangle from activity 6.
☐ Scissors.
☐ A calculator (optional).

GREAT BALL O' FIRE!

1. If you scale down our real earth, moon and sun about 558 *million* times, they have the same *proportions* as the models taped to your wall.

a. How many times bigger are the *real sizes* of the earth, moon and sun than these circles? Write your answer with the correct number of zeros.
b. Are the real earth, moon and sun shaped like these scale models? Explain.
c. Does this model scale the *distances between* the sun, earth and moon as well as their size? Explain.

2. Hold your hand in front of one eye so it looks bigger than the model sun.

MY HAND, BIGGER THAN THE SUN?

a. How did you increase the *apparent size* of your hand? Did its *real size* increase as well? Explain.
b. Why does the model sun appear larger, and the model earth appear smaller, than the real objects outside?

---

3. Cut a strip of adding machine tape (add-tape) 75 cm long. Square off each end.

a. Trace around a nickel to make a row of 10 circles along the upper edge, starting at the left end. The circles should *just* touch, but not overlap.

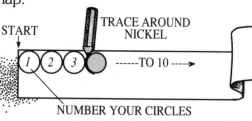

START
TRACE AROUND NICKEL
①②③ -----TO 10 ----->
NUMBER YOUR CIRCLES

b. Mark a 20 nickel and a 30 nickel length without drawing more circles.

①②③④⑤⑥⑦⑧⑨⑩
20 nickels
30 nickels

BE ACCURATE!

c. Lay 10 paper punches across a piece of clear tape held sticky-side-up. Pick them up with a pin so you can place them, just touching, in a straight row.

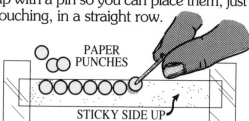

PAPER PUNCHES
STICKY SIDE UP

d. Tape these paper punches to the lower edge of the add-tape beginning at the left end. Number them 1 through 10.

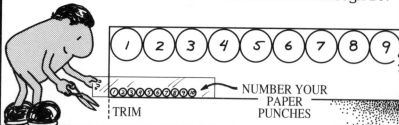

① ② ③ ④ ⑤ ⑥ ⑦ ⑧ ⑨
①②③④⑤⑥⑦⑧⑨⑩
TRIM
NUMBER YOUR PAPER PUNCHES

e. Mark 10-punch lengths, out to 100.

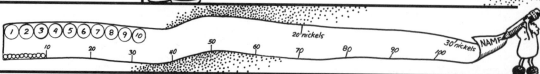

①②③④⑤⑥⑦⑧⑨⑩   10   20   30   40   50   60   70   80   90   100   20 nickels   30 nickels   NAME

---

4. Answer each question using your add-tape ruler and the scale model on the wall: DIAMETER

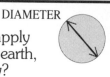

a. How many model earths (nickels) fit in a row across the *diameter* of the model sun?

b. How many model moons (paper punches) fit across the diameter of the model earth?

c. Do your answers apply equally well to our *real* earth, moon and sun? Why?

Objective

To measure the relative diameters of the earth, moon and sun using scale models. To distinguish between apparent size and real size.

Supporting Concepts

○ A 3-dimensional sphere projected straight onto a 2-dimensional plane (a piece of paper) forms a circle. Demonstrate this by casting the shadow of a tennis ball onto a piece of paper held perpendicular to a bright, distant light source (the sun).

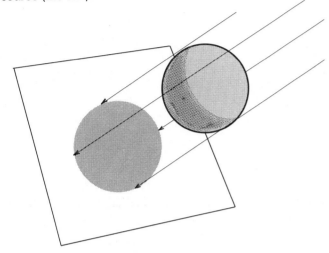

Lesson Notes

1. Prepare these models in advance. See the *Materials* on this page for construction details.

3a. Ideally, the pencil lines (from a reasonably sharp pencil) should just touch. These circles will overlap unless a *slight* gap is left as a pencil allowance when the nickel is placed next to the previously-drawn circle. Sloppiness here will multiply in step 3b.

3b. Allow your students to devise their own methods for marking the 20 and 30 nickel divisions, as long as they are reasonably accurate. They might fold the tape, or mark a 10-nickel distance against notebook paper, or use the measuring triangle.

3c. Again, careful placement of these paper circles will ensure a more accurate calibration of the rest of the ruler in step 3e.

3e. The measuring triangle is an ideal tool to use here.

Answers

1a. The real earth, moon and sun are 558,000,000 times bigger than these models.

1b. No. The real earth, moon and sun are 3-dimensional spheres, while these models are 2-dimensional circles.

1c. No. The model moon and earth are taped to the wall less than one model sun diameter away. This is much too close!

2a. To increase the apparent size of your hand, place it nearer your eye. Though the hand now appears bigger, its real size hasn't grown at all.

2b. These models have different apparent sizes than the real objects outside because we are viewing them at different distances: we are standing on the real earth, while viewing the real sun from a great distance.

4a. About 108 earth diameters fit across the model sun. (Those who measured more earths squeezed the nickel circles closer together along their rulers, while those who measured less used rulers with more widely spaced circles.)

4b. About 3.5 moon diameters fit across the model earth.

4c. Yes. These models were all reduced by the same scale (558 million times). Therefore, the ratio of corresponding model diameters is proportional to actual diameters.

model sun/model earth = 108 =
                 actual sun/actual earth

model earth/model moon = 3.5 =
                 actual earth/actual moon

Materials

☐ Model the earth, moon and sun as shown in step 1. Tape them to your classroom wall:

Cut at least four strips of adding machine tape into 240 cm lengths to "build" the sun. Use masking tape to connect the "spokes" into a rough circle. Label it with an index card. (If you have butcher paper and tempera paint, a solid orange circle is even better.)

Use pen or pencil to draw the outline of a nickel on another index card. Label it "earth." Do not tape a real nickel to the card. Its actual diameter is slightly smaller than its outline diameter.

Tape a paper punch (colored for contrast) to a third index card and label it "moon."

☐ Adding machine tape. (Hereafter we will call this "add-tape," in both the worksheets and teaching notes.)
☐ A meter stick (optional).
☐ Scissors.
☐ Masking tape.
☐ The measuring triangle.
☐ A nickel.
☐ A paper punch tool.
☐ Colored paper.
☐ A straight pin.
☐ Clear tape. The kind you can write on is best. Otherwise, number above each paper punch on the add-tape ruler, at the top edge of the clear tape.

PAPER PUNCH MOON

1. This diagram compares the *diameters* of the earth, moon and sun in 4 important ways.

a. Restate each comparison in a complete sentence.

b. Do your previous findings agree with this new information? Explain.

c. Is this a *scale* drawing? Explain.

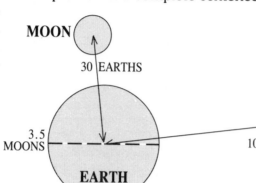

MOON

30 EARTHS

3.5 MOONS

EARTH

108 EARTHS

108 SUNS

SUN

2. Draw a nickel "earth" and a paper punch "moon" on add-tape. Use your ruler to separate them by the correct scale distance.

a. Label the earth, the moon and the distance between.

b. Punch out a hole where you placed the moon.

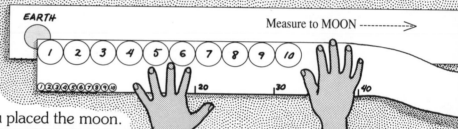

EARTH

Measure to MOON --------->

3. If your model has proportional parts, then the scale moon viewed from the scale earth should have the same *apparent size* as the real moon viewed from our real earth.

a. Imagine that you are standing on your model earth: stretch your model full length with the "earth" at your eye and the "moon" hole against the sky. Do you think this hole has the correct apparent size?

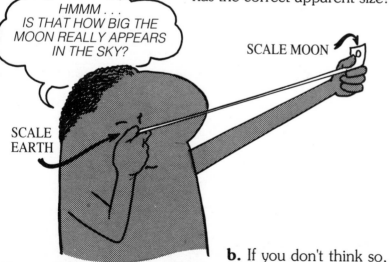

HMMM . . . IS THAT HOW BIG THE MOON REALLY APPEARS IN THE SKY?

SCALE MOON

SCALE EARTH

b. If you don't think so, draw another "moon" beside the hole that looks about right.

4. Check out the accuracy of your model at night against the real moon. What do you find?

a. Ask your family members if they think the paper punch moon is big enough.

b. Record their reactions as part of your answer.

Objective

To develop a model of the earth and moon that correctly scales both size and distance. To confirm that this model correctly predicts the apparent size of the real moon in the real sky.

Lesson Notes

1. These four numbers relationships frame the earth, moon and sun in a scale of comparison that is easy to understand. Because the orbits of the earth and moon are slightly elliptic (not perfectly round), these numbers are only approximate. It is coincidence, not a printing error, that 108 appears twice.

3-4. This model can be viewed from two very different perspectives. Looking at it from the side presents an outside view, representing what an astronaut would see approaching the earth and moon from far away. Looking at it from either end presents an inside view; how an earthling sees the moon, or how a moonling sees the earth.

Answers

1a. 3.5 moon diameters fit across the earth.
108 earth diameters fit across the sun.
30 earth diameters separate the earth and moon.
108 sun diameters separate the earth and sun.

1b. Yes. About 3.5 paper punch "moons" fit across the tracing of a nickel "earth." About 108 nickel tracings fit across the add-tape "sun."

1c. No. Neither the relative sizes of each circle, nor the distances between, are drawn to scale. The sun, for example, is too small compared to the earth and moon, and much too close.

2. Students should use their rulers from the previous activity to mark off a distance of 30 nickel diameters on add-tape, then separate the centers of an outlined nickel and a paper punch by that distance (about 69 cm). The finished model looks like this:

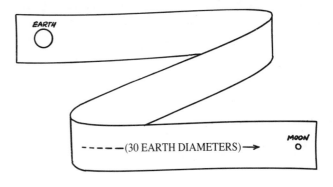

3. Most students will think that the paper punch hole, held an arm's length away, seems much too small to represent the moon in the real sky. They'll draw a moon circle next to it several times larger.

4. When the model is fully extended at about arm's length, the paper punch nicely frames the moon.

4b. Student should report the opinions and reactions of their family members. (Even though we psychologically see a saucer-sized moon in our heads, it has an angular diameter of only 1/2°, the size of a paper punch held at arm's length.)

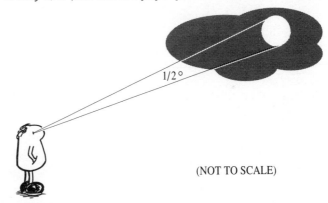

(NOT TO SCALE)

Extension

The moon seems much larger when it's near the horizon than when it's higher in the sky. Use your earth-moon model to find out if this is an illusion. (The moon, in fact, has the same 1/2° angular diameter everywhere in the sky. Its larger apparent size on the horizon is a psychological illusion created by looking across filled space.)

Materials

☐ Add-tape.
☐ A nickel.
☐ A paper punch tool.
☐ The nickel/paper punch ruler from the previous activity.

PINHOLE PROJECTIONS

1. If you scale down the sun about 223 *billion* times, it has the same diameter as a paper punch.

a. Paper punch a hole in the end of a meter of add-tape. Color an orange *corona* around this "sun."

COLORED "SUN"

EARTH?

YOUR RULER

b. Separate it the correct scale distance from "earth." How big should you draw the earth?

c. Label the earth, the sun and the distance between.

2. Lay this earth-sun model next to the earth-moon model you made before.

THESE EARTHS ARE DIFFERENT SIZES . . .

EARTH   EARTH

a. Why is the earth smaller in one model than the other?

b. What do these two models suggest about the *apparent size* of the moon and sun in our sky? Explain your reasoning.

3. Tape two cans together into a hollow tube.

a. Cover one end with waxed paper held by a rubber band.

b. Cover the other end with aluminum foil. Poke a pinhole in the center.

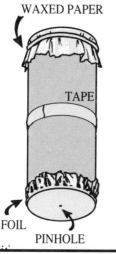

WAXED PAPER

TAPE

FOIL

PINHOLE

4. Cut out the row of tiny <u>Sun</u> <u>Circles</u>. Tape it along the length of the tube.

END TO END

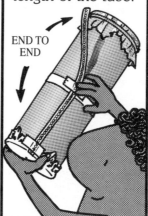

5. Aim the foil end of this tube at the sun so light projects through the pinhole to focus on the waxed paper screen. Shade around the screen with your hands.

FOIL END

SUN PROJECTION

6. A small image of the sun projects through the pinhole onto the screen.

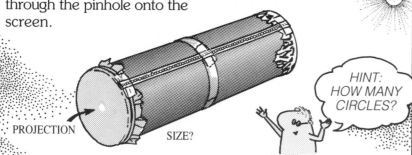

PROJECTION

SIZE?

HINT: HOW MANY CIRCLES?

a. Compare the size of this sun image to the circles on the side of the can.

b. If the pinhole represents the earth, is the projected sun and the distance between properly scaled? Explain.

7. Separate the foil pinhole with enough cans to project a sun image on the screen that is as big as a paper punch. Explain how you did this.

EXTRA CANS . . .

PAPER-PUNCH-SIZED IMAGE

Objective

To develop a model of the earth and sun that correctly scales both size and distance. To verify the scale dimensions of this model with a pinhole projector.

Supporting Concepts

⊕ A pinhole projects real, inverted images on a waxed paper screen, similar to a lens.

• Construct a pinhole projector from two cans as directed in step 3. Ask student volunteers to aim it at a bright distant window or a candle flame, shielding the screen from light as in step 5. Ask what they see. (An inverted image of the distant window scene or candle flame.)

• Explain this phenomenon in terms of a simple blackboard drawing:

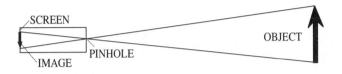

Lesson Notes

1a. Corona means crown. The sun's corona looks like a crown of flame around its perimeter.

5. This projected sun image can also be shaded by holding an additional can, both ends removed, up to the waxed paper screen. This is especially useful for younger children with small hands. (This sun image is safe to view because waxed paper, rather than the eye's retina, absorbs the solar radiation.)

7. This pinhole projector may now be disassembled and the cans saved or recycled. Some students may want to take their projectors home intact.

Answers

1. The finished model is again about 69 cm long, like the last:

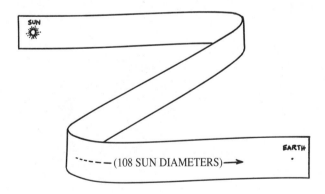

1b. If 108 earths fit across a paper punch sun, the earth can be no larger than a tiny dot.

2a. The earth is smaller in this earth-sun model because it has a smaller scale than the previous earth-moon model. (Here the earth dot and paper punch sun are scaled down 223 *billion* times. The nickel earth and paper-punch moon from activities 8 and 9 are scaled down only 558 *million* times.)

2b. These models suggest that our moon and sun must have the same *apparent* size when viewed from earth. They both look as large as a paper punch when held an arm's length away. (It is difficult to believe that our blazing sun appears no bigger than a paper punch. A full eclipse of the sun by the moon confirms that both bodies have about the same apparent size: an angular diameter of about 1/2°.)

6a. The sun image projected onto waxed paper appears to have the same size as one of the circles taped to the side of the cans.

6b. Yes. The projected sun is separated from the pinhole earth by 108 sun diameters of equal size. (Another scaling requirement is that 108 pinhole earths fit across one of the sun circles. The foil pinhole is clearly too big to fit this many times. A tiny pin prick, barely piercing the foil, would be more to scale. It, too, projects a sun image of identical size. The clarity of the sun image from a smaller hole is in fact, much enhanced, through its intensity is reduced.)

7. Add additional cans (about 6 altogether) to form a tube with the same length as the earth-sun model in step one. Sunlight entering the pinhole "earth" on the foil end projects through a distance of 108 paper-punch suns to form a paper-punch-sized sun image on the waxed paper.

Extension

Try projecting multiple paper-punch sun images onto the waxed paper. Can you demonstrate that each sun has the correct size? (Fit the longer pinhole projector from step 7 with a new piece of foil. Tape a row of 3 paper punches across this foil that *just* touch, then stick a pinhole through the center of each one. These 3 pinholes project 3 sun images on the waxed paper that also *just* touch, proving that all 3 images have the same paper-punch size.)

Materials

☐ A paper punch tool.
☐ An orange crayon or marking pen.
☐ Add-tape.
☐ A meter stick.
☐ Clear tape.
☐ The paper-punch ruler from activity 8.
☐ The earth-moon model from activity 9.
☐ About 7 medium-sized cans per lab group. They should be about 11 cm (4 3/8 inches) tall, of equal diameter and have both ends removed. Ask students in advance to bring these from home. Towel or gift wrap tubes also serve. These will dramatically reduce the number of cans you need to supply. The disadvantage is that narrower tubing must be aimed with greater precision, making the projected sun image more difficult to locate. If you use gift wrap tubes, keep heavy-duty scissors or a sharp knife handy to cut them to length.
☐ Masking tape.
☐ Waxed paper.
☐ A rubber band.
☐ Aluminum foil.
☐ A straight pin.
☐ The Sun Circles cutout.
☐ Direct sunlight. In cloudy weather, complete steps 1 and 2, then proceed to the next activity. Finish up when the weather clears.

PAPER PLATE SUN

1. Cut out the row of small <u>Earth Circles</u>.

a. Get a paper plate. If the diameter of our earth were reduced to the size of one of these circles, the sun would be about as large as this plate. Explain why.

HMMM, HOW MANY "EARTHS" SHOULD FIT ACROSS A PAPER PLATE "SUN"...

STRIP OF "EARTH CIRCLES"

b. Color the paper plate orange and label it "sun."

2. Tape one of your earth circles to the center of an index card.

NOW, HOW MANY "EARTHS" REACH TO THE "MOON?"

a. Draw the moon to scale at the correct distance from the earth. Explain how you did this.

b. Label each circle.

3. Cut a piece of string equal to 9 model suns placed edge to edge. How many times should you lay this string end to end to separate your model sun from its earth and moon? Explain your reasoning.

9 "SUN" DIAMETERS.

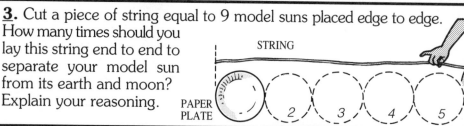

STRING

PAPER PLATE

2 3 4 5 6 7 8 9

4. Take these materials outside:

- ❂ Paper plate sun, index card earth and moon.
- ❂ Measuring string (9 plates long).
- ❂ Add-tape model of earth and sun.
- ❂ This worksheet.
- ❂ Pencil and paper.
- ❂ Book or clipboard to write on.

a. Set your paper plate "sun" on edge (against a wall?). Use your string to place the index-card earth and moon the correct distance away.

b. Does your add-tape model confirm that the paper plate sun has the correct apparent size? Explain.

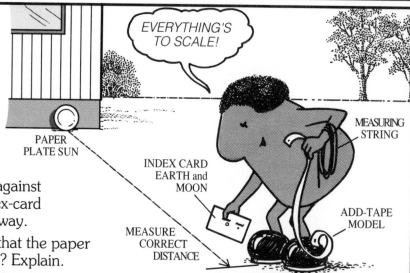

EVERYTHING'S TO SCALE!

PAPER PLATE SUN

INDEX CARD EARTH and MOON

MEASURE CORRECT DISTANCE

MEASURING STRING

ADD-TAPE MODEL

5. Every day the earth orbits *right* (counterclockwise) around its paper plate sun, twice the width of this page.

a. Take a *one week* journey around the sun from where you stand. How far did you move?

b. How long should it take to return full circle?

c. As a space traveler, smaller than a speck of dust, how do you react to this scene? Write a short poem or paragraph to express how you feel.

Objective

To model the sizes of the sun, earth and moon, and the distances between them, all to the same scale. To get a sense of the vast emptiness of space.

Lesson Notes

4-5. If weather conditions are mild, students might answer these questions outside, standing at a distance of 108 plate diameters from the paper-plate sun (27 yards or 25 meters). If the weather is uncomfortable, they should take brief notes, then compose more complete answers back in the classroom.

Answers

1a. About 108 Earth Circles fit across the diameter of a paper plate, from edge to edge, through its center. The diameter of the plate and smaller earth circle, therefore, have the same proportions as the diameter of our real sun and real earth.

2a. Determine how small a moon must be for 3.5 of them to fit across one of the earth circles. Draw one of these tiny moons 30 earth circles away from the one you taped on your card.

2b. The earth-moon system should look like this (actual size) drawing. The moon should be drawn about the size of a pinhole.

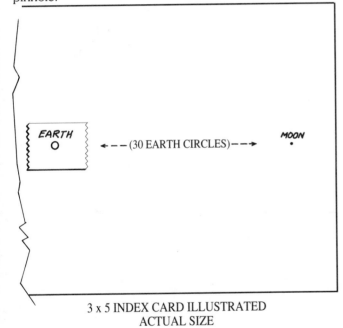

3 x 5 INDEX CARD ILLUSTRATED
ACTUAL SIZE

3. The earth and sun are separated by 108 sun diameters. These models, therefore, should be separated by 108 paper plate diameters. If the string is cut 9 plate diameters long, then laying the string end to end 12 times covers the required distance: 9 plate diameters x 12 = 108 plate diameters.

4b. Yes. At a distance of 108 plate diameters, when you extend the add-tape model with the earth dot at your eye, the distant paper-plate sun appears perfectly framed inside the paper punch. Both models predict the same apparent sun size.

5a. In 1 week, move 14 paper-widths to the right.

5b. It takes one year for the earth to orbit the sun.

5c. Students might express their feelings about these or other topics: earth is tiny; space is vast; the sun is very distant compared to the moon; the sun must produce vast energy to warm us from so far away; we receive only a tiny fraction of its total energy.

Extension

Show that the earth really moves through a distance of 2 worksheet widths every 24 hours:

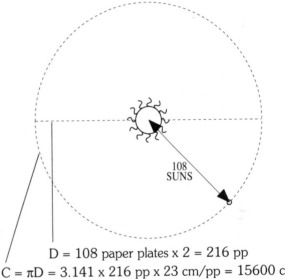

108
SUNS

D = 108 paper plates x 2 = 216 pp
C = πD = 3.141 x 216 pp x 23 cm/pp = 15600 cm
15600 cm/year x 1 year/365 days = 42.7 cm/day
This is about twice the length of a 21.5 cm worksheet.

Materials

☐ The Earth Circles cutout.
☐ Scissors.
☐ A generic paper plate, 9 inches (23 cm) in diameter. See teaching notes 20.
☐ An orange crayon or marking pen. Coloring the paper plate is optional. No more than one plate per lab group need be colored.
☐ Clear tape.
☐ An index card.
☐ String.
☐ The add-tape model of the earth and sun from the previous activity.
☐ A clipboard, book, or other portable writing surface.
☐ A non-rainy day.

ECLIPSE

1. Set a tennis ball between two batteries. Find its *diameter* by measuring the distance between the batteries with your measuring triangle.

BALL DIAMETER

2. If this tennis ball represents the earth, find the diameter of the moon at the same scale. Use batteries to measure a ball of clay to the size you calculated.

MY FIGURES !

TENNIS BALL EARTH

CLAY MOON

3. Prop a flashlight "sun" on clay so it shines straight onto a paper screen about 60 cm away. Tie a string to a pin stuck in your clay ball.

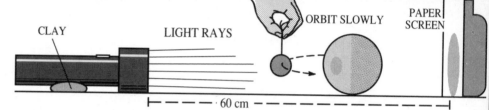

CLAY LIGHT RAYS ORBIT SLOWLY PAPER SCREEN

|← ――――― · 60 cm ―――――― →|

a. Orbit your "moon" around the tennis ball "earth," just in front of the paper screen, so you can see the shadows.

b. In an *eclipse of the sun*, the moon casts its shadow across the earth. Model this event, then draw a labeled diagram.

c. In an *eclipse of the moon*, the earth casts its shadow across the moon. Model this event, then draw a labeled diagram.

4. Stick a pin halfway into your earth. Protect your finger with 3 short layers of masking tape.

*THIS STRING SHOULD MODEL **DISTANCE** IN THE CORRECT SCALE.*

a. Tie the pins on your earth and moon together, using just enough string to separate them by the correct scale distance. Show your calculations.

b. Use your earth-and-moon tape (from activity 9) to confirm that this model is scaled correctly. Explain how you did this.

5. Model an eclipse of the sun again, this time in direct sunlight at the correct scale distance.

SUNLIGHT

a. Can the moon's shadow cover the whole earth?

b. Can every observer on the lighted side of the earth see this eclipse? Explain.

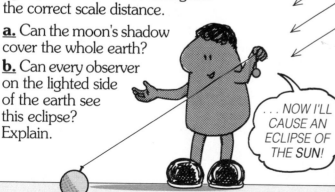

. . . NOW I'LL CAUSE AN ECLIPSE OF THE SUN!

6. Now model an eclipse of the moon.
a. Can the earth's shadow cover all the moon?
b. Can every observer on the dark side of the earth see this eclipse? Explain.

7. Your earth and moon models have so far been used with a flashlight "sun" and the real sun, but not with a sun modeled to the correct scale.

a. How big should this scale sun be compared to a tennis-ball earth? Is this bigger than your room?

b. How far away should this scale sun be? Is this more than a mile? (1 mile = 1,600 meters)

Objective

To build a scale model of the earth and moon using a tennis ball. To model eclipses of the sun and moon.

Supporting Concepts

⊙ Light travels in straight lines. This property is represented in drawings as light "rays." Shadows are caused by the absence of light.

⊙ A tree eclipses (hides) the sun when you stand in its shadow. If you don't stand in its shade, you won't see this eclipse.

⊙ A tree's shadow eclipses (hides) your body as well. Walk from direct sunlight into shade, and you are less visible.

Teaching Notes

1. The tennis ball sticks up higher than the batteries, and thus gets in the way of measuring. Simply roll it out from between the batteries after they are properly spaced.

2. Students should use their measuring triangles to space the batteries a scale-moon diameter apart, then roll a ball of clay that just fits between.

The diameter of this clay moon is 3.5 times smaller than the diameter of the tennis ball. Yet it seems much smaller than this! While the diameter of the clay ball (a linear dimension) is reduced 3.5 times, its projected area (a circle) is 3.5^2 times smaller, and its volume (what we see) is 3.5^3 times smaller. You can illustrate this by cutting a tennis ball in half. At the end of the lesson, collect all the clay moons your students have made. Find how many you can jam into one hemisphere. You should count $3.5^3/2$, or about 21 moons.

3. Notebook paper will serve as a screen. Students might fix it to a bottle or cereal box with a piece of rolled tape. As they orbit the clay moon around the earth, it casts a small shadow onto this screen that disappears into, and emerges from, the larger shadow of the earth. This allows an eclipse of the moon to be detected under fairly bright conditions. In a dark room, this background screen is not required.

5. This step is most easily carried out while standing on a smooth, light surface, such as concrete. It shows the small, elusive moon shadow much better than black asphalt or irregular grass.

Answers

1. Diameter of tennis ball = 6.5 cm.

2. The moon's diameter is about 3.5 times smaller than earth's diameter:

moon diameter = 6.5 cm/3.5 = 1.9 cm

3b. An eclipse of the sun by the moon:

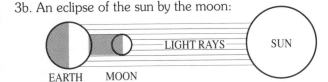

3c. An eclipse of the moon by earth's shadow:

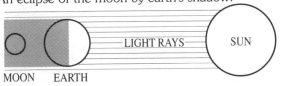

4a. The earth and moon are separated by a distance of about 30 earth diameters:

length of string = 6.5 cm x 30 = 195 cm

(Students should add extra length to allow tying at each end. Those who don't plan ahead might join the string to each pin with tape.)

4b. Hold the "nickel" earth next to the tennis ball earth. Extend the moons on both models to their full length. Sight from the earth(s) to both moons. The paper-punch moon will perfectly frame the clay moon, as it frames the real moon in the night sky.

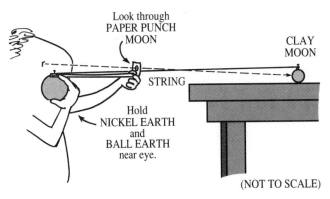

5a. No. The moon's shadow covers only a small portion of earth's lighted side.

5b. No. Only those observers standing in the moon's shadow will see that it covers (or partly covers) the sun.

6a. Yes. The entire moon can be covered by the earth's shadow.

6b. Yes. The moon is visible everywhere on the dark side of the earth, so its eclipse can be seen by any observer.

7a. The earth's diameter fits into the sun's diameter about 108 times:

sun diameter = 6.5 cm x 108 = 702 cm = 7.02 m

Students should measure off 7 meters, then try to imagine a sphere of this diameter fitting into their room.

7b. The earth and sun are separated by a distance of about 108 sun diameters:

earth-sun separation = 7.02 m x 108 = 758 meters

This is slightly less than 800 meters, or 1/2 mile:

758 m x 1 mile/1600 m = 0.48 miles

Materials

☐ A tennis ball, new or used. Be aware that once you pierce it with pins, the ball will lose much of its bounce.
☐ Two batteries, dead or alive. Size D work best.
☐ The measuring triangle.
☐ Clay.
☐ A calculator (optional).
☐ A flashlight, string and scissors.
☐ Straight pins.
☐ A meter stick.
☐ Notebook paper.
☐ A bottle, cereal box or wall to support the notebook paper.
☐ Masking tape.
☐ Direct sunlight. If the weather is cloudy, skip steps 5-7 plus the next activity and continue with lesson 14. Return on a sunny day.

MOON PHASES

1. Cut around the long black rectangle labeled <u>Moon Phase Tabs</u>.

2. Fix a Ping Pong ball to the end of a straw with clear tape.

3. Take these materials to a place that receives direct sunlight. Hold the Ping Pong "moon" in front of you while *slowly* turning your body *counterclockwise*:

a. Watch the sunlit area of your moon:

When does it grow?

When does it shrink?

b. Starting with the fully shadowed *new moon* tab as number 1, correctly order the remaining moon phase tabs by numbering each white box.

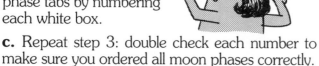

c. Repeat step 3: double check each number to make sure you ordered all moon phases correctly.

4. Cut out the <u>Sunlight Circle</u>. Fix it to the middle of a paper plate with rolled tape.

ROLLED TAPE

a. Cut out each moon phase tab. Tape these around the edge of the circle in the correct order. Begin with the new moon.

b. Remove the moon ball from the straw. Cover precisely half of it with black tape. (Follow the seam line of the ball.)

CENTER SEAM

c. Fix it to the center spot on the plate so its light side faces directly into the "sunlight." Do this by taping a paper clip "foot" to the ball, then tape this "foot" to the plate.

d. Fold each tape hinge forward so the tabs stand straight up.

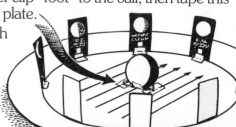

5. Set your model near *eye level* with space to walk full circle around it. A chair on a desk might work, or ask a lab partner to hold it.

a. Imagine you are in a spaceship circling the moon *counterclockwise*. List all the phases you see on your moon model in the correct order.

b. Now circle the moon in a *clockwise* direction. What difference does this make?

c. How much of the moon's surface is in sunlight at any one time?

Objective

To model the phases of the moon. To describe its cycle of phases in the correct order.

Supporting Concepts

❶ The hands on a clock are said to move clockwise. Counterclockwise describes circular motion in the opposite direction:

• Ask students to imagine themselves standing on a clock face. Direct them to turn clockwise or counterclockwise.

• To open a jar of peanut butter, which way should you turn the lid? (counterclockwise)

• To tighten a screw, which way should you turn the screwdriver? (clockwise)

Lesson Notes

3. To see the moon phases pictured on the tabs, you must orbit the Ping Pong "moon" in approximately the same plane that the sun and your head are in. If the sun is relatively low on the horizon, this plane is nearly parallel to the ground. Higher altitudes of the sun, however, require a compensatory tilt: sweep the ball lower through its full moon phase, higher through its new moon phase.

The plane of the moon's orbit and the earth-sun plane (known as the ecliptic) don't quite coincide. The angle between them is 5°. This difference explains why the earth's shadow eclipses the moon only on rare occasions, not at every full moon. Unfortunately, your head *can* eclipse the Ping Pong ball at every full moon. It is too close and too large to represent the earth to scale. To view the full moon, compensate by lifting the ball's orbit above your head's shadow.

3b-c. All 8 moon phases must be correctly ordered, *before* the tabs are cut and taped into position in step 4.

4a. Notice that each piece of tape functions like a hinge. It must be centered at the bottom of each tab as shown, so the tab remains free to pivot into an upright position.

4b. Black electrical tape tends to lose its grip over time. To minimize lifting, snip into the free edge of the first strip you wrap around the ball's seam. Then secure with cross pieces.

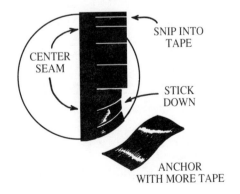

CENTER SEAM

SNIP INTO TAPE

STICK DOWN

ANCHOR WITH MORE TAPE

5. The black and white Ping Pong "moon" must be viewed at eye level. In this special position it looks just like the moon phase tab directly behind it.

The cusps, or horns, of the crescent moon always point away from the sun like the curve of a bow, with its imaginary arrow aimed at the sun. (This arrow lies roughly within the ecliptic, the earth-sun plane. Because the paper plate models a horizontal ecliptic, the crescent moon's imaginary arrow aims level as well. The crescent moon in the real sky, however, may tilt with respect to the observer's horizon, as this horizon also tilts with respect to the ecliptic.)

Answers

3a. The sunlit portion of the moon grows as you turn it away from the sun. It shrinks as you turn it toward the sun.

3b. The correctly numbered strip looks like this:

WAXING CRESCENT	THIRD QUARTER	FULL MOON	FIRST QUARTER	WANING CRESCENT	WAXING GIBBOUS	NEW MOON	WANING GIBBOUS
2	7	5	3	8	4	1	6

5a.
1. new moon
2. waxing crescent
3. first quarter
4. waxing gibbous
5. full moon
6. waning gibbous
7. third quarter
8. waning crescent

5b. When you circle the moon clockwise, you see the same moon phases, but they happen in the opposite order.

5c. Exactly half of the moon is illuminated by sunlight at all times (except during an eclipse). We see more or less of this lighted side depending on our position relative to the moon and sun.

Materials

☐ The cutout sheet for this activity with Moon Phase Tabs and Sunlight Circle. Supply one sheet per activity group or student pair.

☐ Scissors.

☐ A Ping Pong ball.

☐ Clear tape.

☐ A straw.

☐ Direct sunlight. A low altitude sun is somewhat easier to use than a high altitude sun. If your schedule is flexible, you might schedule science to happen earlier or later in the day. In cloudy weather skip ahead to the next activity. If the weather refuses to clear, a slide projector light makes a reasonable indoor substitute. Position it at a distance to minimize eclipsing the ball in the shadow of your head.

☐ A paper plate and paper clip.

☐ Black tape. Any flexible cloth, vinyl or electrical tape will work.

WHERE YOU LIVE

1. Trim across the top and bottom of the <u>Globe</u> <u>Gores</u>. Carefully cut out these *gores*, keeping them joined at the equator like a chain.

TRIM **TOP** AND **BOTTOM** FIRST

TAKE TIME TO DO A GOOD JOB

2. Wrap these gores around a tennis ball. Join the equator into a circle with a bit of clear tape.

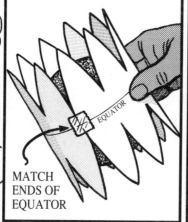

EQUATOR

MATCH ENDS OF EQUATOR

3. Set the tennis ball on a small jar, with the gores pointing up and down.

a. Tape an opposite pair of gores so their points *just* touch. Turn the ball over. Tape the *same* pair to just touch on the other side.

SMALL STRIPS OF TAPE.

MATCH OPPOSITE PAIRS

SMALL JAR

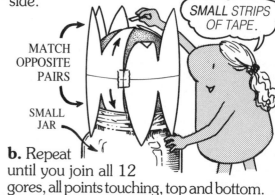

b. Repeat until you join all 12 gores, all points touching, top and bottom.

4. Pad your finger with 3 short layers of masking tape. Push a pin straight into the ball at each pole (where the gores meet) so it sticks straight out about 1/2 cm.

$\mathbb{I}$ 1/2 cm: ACTUAL SIZE

NORTH POLE

ARCTIC CIRCLE

EQUATOR

ANTARCTIC CIRCLE

SOUTH POLE

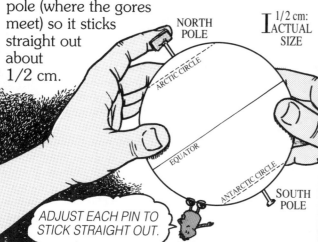

ADJUST EACH PIN TO STICK STRAIGHT OUT.

TAPE

N E S

5. Punch out the <u>Mini</u> <u>Compass</u>. Stick a small square of clear tape over the top, then push a pin through its center.

a. Push this pin into the ball to show where you live. Make it stick straight out about 1/2 cm.

b. Align **N** on your tiny "compass" to the north pole of your globe, then stick it in place.

WHERE I LIVE . . .

N E S W

EQUATOR

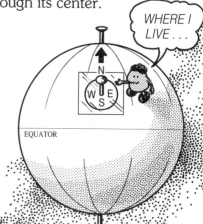

6. Find out what's on your globe:

<u>**a.**</u> List 7 continents and 3 major oceans.

<u>**b.**</u> Find 0° *longitude*. List all degrees of longitude you find, going west full circle around the *equator*: 0°, 20°W, 40°W, ...

<u>**c.**</u> Find 0° *latitude*. List all degrees of latitude you find, going north full circle around the *prime meridian*: 0°, 20°N, 40°N, ...

<u>**d.**</u> Estimate the longitude and latitude of where you live. Repeat for a point directly opposite your home, on the other "side" of the earth.

<u>**e.**</u> Name 5 earth "belts" located at these latitudes: 0°, 23.5°N, 23.5°S, 66.5°N, 66.5°S.

<u>**f.**</u> The *ecliptic circle* is broken into how many separate pieces?

7. Suppose an apple falls to the ground from a tree in South America, Japan and Australia at the same time.

a. Draw a round earth. Show the *direction* that each apple falls to earth.

b. Redefine *up* and *down* in terms of *in* and *out*.

Objective

To model the earth with a tennis ball. To identify its major geographic features and standard lines of reference.

Supporting Concepts

❑ Construct a tennis ball "earth" in advance. Show your class how mapmakers customarily divide it into lines of longitude and latitude:

• Lines of longitude run north-south from pole to pole. (<u>Long</u>itudes are always <u>long</u>.)

• Lines of latitude run east-west, parallel to (and including) the equator. (Latitudes are sometimes short.)

• Both longitude and latitude are measured in degrees from earth's center, similar to the azimuth circle on your compass.

Lesson Notes

1. This cutting operation requires 10 to 15 minutes of concentrated effort. Caution students to take their time and do a good job.

2-3. Younger students typically use too much tape. Narrow snippets cut *almost* across the width of a strip of tape are easy to handle and make an attractive presentation. Dispense these, about 15 pieces per student, stuck to any suitable object.

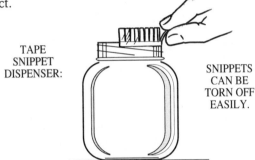

TAPE SNIPPET DISPENSER:

SNIPPETS CAN BE TORN OFF EASILY.

3. Tape each opposite pair of gores, both top and bottom, before continuing to the next pair. This assures that all remaining gores remain centered over the equator. If the ball you are covering is new and fuzzy, you will need to pull firmly on the gores to make them meet at a common point. If you are covering a well-worn ball, allow a looser fit, so the gore points touch without overlap.

4-5. Pushing pins into tennis balls is hard on the skin. Keep tape protection on your finger through both of these steps until all 3 pins are placed. Alternately, protect your finger with several layers of clothing.

6. Mapmakers choose to show only those imaginary lines of longitude and latitude they deem appropriate. Our globe is calibrated in increments of 10°, but only lines of astronomical significance are drawn in.

Answers

6a. North America   Pacific Ocean
     South America   Atlantic Ocean
     Europe   Indian Ocean
     Africa
     Asia
     Australia
     Antarctica

6b. 0°, 20° W, 40° W, 60° W, 80° W, 100° W, 120° W, 140° W, 160° W, 180°, 160° E, 140° E, 120° E, 100° E, 80° E, 60° E, 40° E, 20° E, 0°.

6c. 0°, 20° N, 40° N, 60° N, 80° N, (over pole), 80° N, 60° N, 40° N, 20° N, 0°, 20° S, 40° S, 60° S, 80° S, (over pole), 80° S, 60° S, 40° S, 20° S, 0°.

6d. We TOPS folk live near Portland, Oregon at 122° W longitude and 45° N latitude. The point opposite our home is 58° E (180° – 122°) and 45° S. (North America is not opposite China, as many people think. It lies opposite South Africa and Australia, in the southern Indian Ocean. To be opposite China, you would have to live in South America.)

6e. 0°   equator
     23.5° N   Tropic of Cancer
     23.5° S   Tropic of Capricorn
     66.5° N   Arctic Circle
     66.5° S   Antarctic Circle

6f. The ecliptic circle on our globe is broken into 12 separate pieces, one for each month of the year. (The location of these months on the ecliptic circle mark the *latitude* where the sun passes straight overhead at that time of year. This concept will be developed in activity 17.)

7a.

JAPAN

SOUTH AMERICA

AUSTRALIA

7b. Up always means *out* from earth's center. Down always means *in*.

Materials

❑ The cutout sheet for this activity with Globe Gores and the Mini-Compass. Though students can use their earth model in pairs, each individual may want to build one.
❑ Scissors. Supply sharp, good quality scissors if available. They will make the cutting task easier, more precise and more enjoyable.
❑ Clear tape.
❑ A tennis ball, new or used.
❑ A baby food jar.
❑ Masking tape.
❑ Straight pins.
❑ A paper punch tool. Or you can cut out the Mini-Compass with scissors.

SOLAR EXPOSURE

1. Rubber band a loop of string to the mouth of a bottle, as tightly as you can.

TIGHT RUBBER BAND!

LOOP

a. Pull up enough string to make a big loop. Fold the top half back and down into "mouse ears."

PULL UP, FOLD OVER

b. Close these "ears" like a book, to form a noose.

HOLD THIS DOUBLE LOOP OPEN

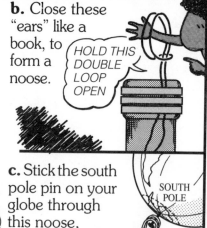

c. Stick the south pole pin on your globe through this noose, and pull it tight.

SOUTH POLE

PULL TIGHT

THROUGH DOUBLE LOOP

2. Pull down on the ends of the string to hold the pin securely against the lip of the bottle, with the globe resting on top.

3. Cut off the spout side of a cardboard milk carton.

4. Add gravel so it supports your bottle at a steep angle. The globe won't fall off the bottle if you keep the south pole pin facing up.

SOUTH POLE ABOVE BOTTLE

GRAVEL

5. Wrap a tape "handle" around a short segment of straw. Take this, plus your compass and your leaning globe, into direct sunlight.

SUN RAYS

TAPE

STRAW

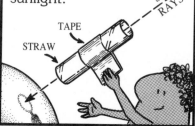

6. Adjust the globe so your "home pin" points straight up and the north pole pin leans due north. You now have an astronaut's eye-view of planet earth!

a. Compare your shadow on the real earth with the pin's shadow on its model earth. Does the sun occupy the same position in both worlds? Explain.

YOUR HOME PIN

NORTH POLE

← - - - NORTH

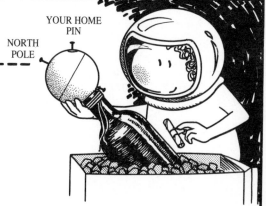

b. Use your straw to find where on earth, right now, the sun appears straight overhead. Explain how you did this.

c. Explore the ring of twilight around your globe (where night meets day). Where on earth is it now twilight? Where do observers in this ring see the sun?

d. The sun appears to "rise" in the east and "sink" in the west. In which direction must you turn your globe to make this happen?

e. Why do we experience day and night? Are days and nights now equal near the earth's poles? Explain.

f. Imagine yourself standing on a giant tennis ball, turning under a fixed sun. Write a poem.

Objective

To model how the earth looks from space. To account for the sun's apparent motion around the earth.

Lesson Notes

1-2. A tightly stretched rubber band firmly holds the string in place against the neck of the bottle. Younger students may not be able to wind it with enough tension. Provide assistance as necessary, or compensate by directing them to add a second rubber band directly below the first.

The noose, when properly formed, allows free rotation of the globe without twisting the string or winding it around the pin shaft. Yet it anchors the globe firmly in place at the bottle's rim. Push the globe off the rim, and it simply hangs at the side of the bottle, held fast by the pin head.

With continued manipulation of the globe, its string may gradually loosen, causing the anchor point to wobble. As this happens, simply pull the string tight again. If necessary, rewind the rubber band with more tension.

4. You must tilt the bottle rather steeply (as illustrated) for northerly latitudes to rotate to a "straight up" position. Laying the bottle flat on its gravel bed raises even "down-and-under" Australia to the "top" of the globe.

Answers

6a. Students should report that their body shadow azimuth (measured by the compass) equals the pin shadow azimuth (cast across the mini compass.) Because the sun lies 180° opposite any shadow it casts, its azimuth must be the same in both worlds. Advanced students might further notice that the length of each shadow is proportional to the body that cast it. This suggest that sun rays strike both worlds from the same altitude.

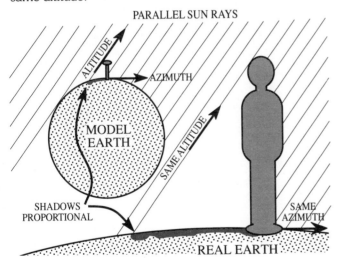

PARALLEL SUN RAYS

ALTITUDE

AZIMUTH

MODEL EARTH

SAME ALTITUDE

SHADOWS PROPORTIONAL

SAME AZIMUTH

REAL EARTH

6b. The sun appears straight overhead at the center of the lighted portion of the globe. (Students should report this location either in terms of longitude and latitude, or with reference to geographic landmarks. To find this overhead position experimentally, move the straw over the surface of the globe, as illustrated in step 5, with its length always aimed toward the globe's center. The sun is directly overhead where the straw casts a thin, round shadow, around a bright center, onto the globe's surface.)

6c. Students should report the sun's position at various geographic locations around this ring of twilight. To any observer on this twilight ring, the sun appears low on the horizon toward the ring's center. On the northeastern part of this ring, for example, the sun appears low on the observer's southwestern horizon. (Students can confirm this with the straw. To allow light to shine through, it must be aimed low, at a near tangent to the globe.)

6d. Turn the globe from west to east. Looking down on the north pole from above, this means turning the globe counterclockwise.

6e. We experience day and night because the earth rotates on its axis.

If your calendar date is currently near the beginning of spring or fall, your students will see that the ring of twilight "touches" each pole. All points in the polar region (and over the entire globe) turn through equal portions of shaded night and illuminated day.

At other times of the year, especially near the beginning of winter or summer, your students will notice that the ring of twilight misses both poles, exposing one (the summer pole) to continuous light and the other (the winter pole) to continuous shadow. Regions near each pole (and over the entire globe) turn through unequal portions of shadow and sunlight.

6f. This step offers your students the possibility of breaking through the very powerful illusion that the sun circles the earth.

Materials

☐ A rubber band. Thick ones work best. If you have only thin ones, apply several.

☐ Kite string or thread. Use heavy-duty (button and carpet) thread if available. Dental floss is also suitable.

☐ Scissors.

☐ The tennis ball globe previously constructed.

☐ A glass pop or beer bottle. Reserve a second bottle of equal height to support a model sun in the next activity. Assign students to work together in pairs to conserve materials. (Groups larger than 2 will reduce hands-on learning opportunity.)

☐ A cardboard milk carton. Half gallon or quart sizes are both suitable.

☐ Clean gravel or sand.

☐ Masking tape.

☐ A straw.

☐ The compass.

☐ Direct sunlight. If the weather is cloudy, ask students to complete step 2, then proceed to the next activity. They should finish this activity on a sunny day.

EARTH DANCE

1. Wrap masking tape around the mouth of a bottle so half its width rises above the glass. Cut this exposed tape into many vertical strips.

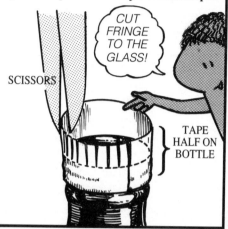

CUT FRINGE TO THE GLASS!

SCISSORS

TAPE HALF ON BOTTLE

2. Fan this fringe outward. Stick a tennis ball "sun" on top.

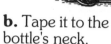

"SUN"

a. Pull a paper clip to a right angle.

b. Tape it to the bottle's neck.

b. Tape a similar paper clip to the neck of your globe bottle, directly under the south pole pin.

EARTH GLOBE

SOUTH POLE PIN

3. The earth *rotates* on its axis while it *revolves* around the sun. Both motions happen counterclockwise.

a. Practice these motions. As you move the models, *always* keep both paper clips pointing in the same direction.

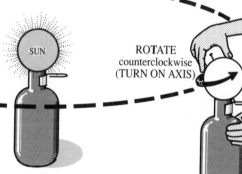

REVOLVE counterclockwise (ORBIT AROUND SUN)

PAPER CLIPS ALWAYS ALIGNED

SUN

ROTATE counterclockwise (TURN ON AXIS)

EARTH

b. Which earth motion causes day and night?
How long does it take the real earth to complete 1 full cycle?

c. Which earth motion causes the tilted north pole to sometimes lean *toward* the sun, and sometimes *away*?
How long does it take the real earth to complete 1 full cycle?

d. Account for earth's changing seasons.

e. How many rotations of the earth happen during each revolution around the sun?

f. Is this working model of the earth and sun also a *scale* model? Explain.

4. Cut out the <u>Concept List</u> for this activity. Work together on this list with a friend:

a. YOU: Choose *any* concept. Use your models to show your friend why the concept is true.

b. FRIEND: Write your initials in the box to certify that you *both* understand the explanation.

NOW I'LL DO THIS ONE: "WINTER NIGHTS ARE MUCH LONGER..."

c. Trade roles and try a new concept.

d. How many concepts did you both understand?

Objective

To model earth's rotation on its axis and revolution around the sun. To understand why night follows day and season follows season.

Supporting Concepts

✪ Demonstrate these vocabulary words with a tennis ball globe. Ask your students to follow along using their own globes:

• axis: This is an imaginary line that runs through the center of the globe, connecting both poles.

• rotate: Turn the globe on its axis, on top of its bottle support. The bold "**t**" in ro**t**ate suggests an axis of ro**t**ation.

• revolve: Orbit the globe *and* its bottle support around a central point on your desk. The bold "**o**" in rev**o**lve suggests a circle of rev**o**lution.

• hemisphere: Point out earth's two half-spheres separated by the equator. Which continents lie mostly within the southern hemisphere? (Antarctica, Australia and South America)

Lesson Notes

2. This sticky fringe may not adhere very well to newer, fuzzier tennis balls. If it lets go, reinforce with extra pieces of tape stuck crosswise over the fringe.

3a. This model suggests (correctly) that the axis of the earth tilts relative to the earth-sun plane. These paper clips, when pointing in the same direction, keep the earth tilting in the proper seasonal orientation with respect to the sun. Remind students, as necessary, to always keep the paper clips aligned as they use these models.

Answers

3b. Earth's ro**t**ation on its axis causes day and night. A complete ro**t**ation happens once every 24 hours.

3c. Earth's rev**o**lution around the sun causes the tilted north pole to lean in different directions relative to the sun at different times of the year. It takes 1 year to complete a full rev**o**lution around the sun.

3d. Summer happens in the northern hemisphere when earth's north pole tilts toward the sun. This bathes the northern hemisphere in sunlight that is more direct, for longer periods of time. As the earth continues to rev**o**lve, its north pole tilts sideways relative to the sun (fall), then away (winter), then sideways (spring), and finally toward the sun for another season of summer. (Seasons in northern and southern latitudes occur 6 months apart.)

3e. About 365 ro**t**ations of the earth happen during each rev**o**lution around the sun.

3f. No. The sun is much too small compared to the earth, and much too close.

4. Some of these explanations are complex, but the challenge can stimulate real "aha!" experiences. If necessary, ask class volunteers to explain the harder concepts, or demonstrate them yourself. Step 4d is the only needed written response in this otherwise oral exercise.

It gets light and dark once a day: Your home pin ro**t**ates into the sun (day) then away from the sun (night).

The sun appears to rise in the east and set in the west: Your home pin ro**t**ates counterclockwise (toward the east),

into a morning sunrise and away from an evening sunset.

Each year, there is a warm summer season and a cold winter season: As earth rev**o**lves, its north pole tilts toward the sun during northern summer, away during northern winter.

The earth's warm and cold seasons are separated by a milder spring and fall: As earth rev**o**lves between summer and winter, its axis tilts sideways relative to the sun, casting equal light on the northern and southern hemispheres.

Summer in one hemisphere is winter in the other hemisphere: While the north pole tilts toward a summer sun, the south pole tilts away from a winter sun.

Winter and summer happen once in a year: Earth's tilted north pole leans once toward the sun and once away during each rev**o**lution.

Winter nights are longer than summer nights: During northern winter (in the U.S. and Canada, for example), your home pin tilts away from the sun. As earth ro**t**ates, this pin turns through a relatively short arc of day and longer arc of night. (This question nicely ties in azimuth observations for the rising and setting sun taken back in activity 4. In winter, the home pin tilts into a southeastern sunrise and away from a southwestern sunset. In summer the sun rises and sets to the north of due west.)

Your noontime shadow is shortest in summer, longest in winter: In the U.S. and Canada, your home pin tilts toward the summer sun: shadows are relatively short because the sun appears high in the south. But your home pin tilts away from the winter sun: shadows are relatively long because the sun appears low in the south.

Your shadow at high noon always points north in Canada, south in Southern Australia: In Canada, the sun always appears to the south and casts a northerly shadow. In Southern Australia, the sun always appears to the north and casts a southerly shadow.

Winter nights are much longer near the poles than at the equator: During winter, earth's north pole tilts away from the sun into nearly 6 months of continuous night. As you approach this dark pole, you experience shorter and shorter periods of daylight.

At earth's poles, the sun circles nearly parallel to your horizon, rising and setting just once a year: At either pole, the earth turns about a fixed point. This makes the sun appear to turn in circles, above the horizon in summer, below the horizon in winter, rising in the spring and setting in the fall.

Twilight near the poles is much longer than twilight near the equator: The sun rises and sets near the poles at low angles, relatively parallel to the horizon, causing long twilights. The sun rises and sets near the equator at steeper angles, nearly perpendicular to the horizon, so twilights are brief.

Materials

☐ A glass bottle with the same height as the one supporting the earth globe.
☐ Masking tape.
☐ Scissors.
☐ A tennis ball.
☐ Two paper clips.
☐ The earth globe on its supporting bottle.
☐ The Concept List cutout.

THE ECLIPTIC PLANE

1. Snip open a short length of straw and cut off 4 small "beads."

CUT:

BEAD:

2. Space them on a rubber band "necklace" around the middle of your tennis ball "sun" Use a canning ring to make them level.

LEVEL

3. Cut around the Ecliptic Ring, inside and out. Rest it on the straw "beads" so the arrow at JUN points in the same direction as the paper clip below. Hold this position with a small hinge of tape.

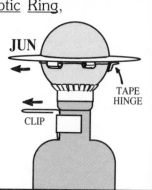

JUN

TAPE HINGE

CLIP

4. Imagine that this ecliptic ring extends outward in all directions. This imaginary "surface" is called the *ecliptic plane*. This plane passes through the center of the sun and earth.

a. Set your model earth next to the sun. What *fraction* of the earth extends above the ecliptic plane? What fraction of the sun extends below it?

b. Does your home ever pass through this ecliptic plane? Explain.

c. Revolve your model earth around the sun, passing each month (on the ecliptic ring) in order. List the four seasons that you see on the ring.

d. Name the special day that begins each new season.

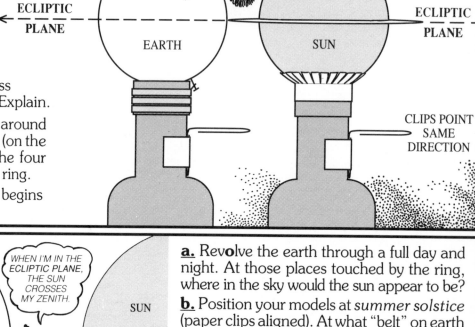

ECLIPTIC PLANE PASSES THROUGH CENTERS!

ECLIPTIC PLANE

EARTH

SUN

ECLIPTIC PLANE

CLIPS POINT SAME DIRECTION

5. Bring your models close together (paper clips aligned) so the sun's ecliptic ring touches the earth.

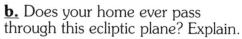

WHEN I'M IN THE ECLIPTIC PLANE, THE SUN CROSSES MY ZENITH.

ECLIPTIC PLANE

SUN

EARTH

a. Revolve the earth through a full day and night. At those places touched by the ring, where in the sky would the sun appear to be?

b. Position your models at *summer solstice* (paper clips aligned). At what "belt" on earth does the sun appear straight overhead as you rotate the globe?

c. Revolve the earth around the sun to the start of each new season. At what "belt" is the sun overhead during these special times of the year?

"BELTS"

6. The *ecliptic circle* printed on your globe is divided into 12 separate month lines. Through what range of latitudes does the sun move in MAR? In APR? In MAY? In JUN?

7. Get the Concept List for this activity. Work with a friend as before. How many concepts did you both understand?

Objective

To model the ecliptic plane. To trace the apparent northward and southward movements of the sun in terms of this plane.

Supporting Concepts

Pantomime these different dimensions.
- *0 dimensions:* point to an imaginary *point* floating in space.
- *1 dimension:* stretch an imaginary *line* between your hands, like a string. Ask two volunteers to stretch it across your room.
- *2 dimensions:* move the palms of your hands over an imaginary *plane.* Ask volunteers to form planes that extend vertically, horizontally, and in other random directions.
- *3 dimensions:* encompass a large *sphere* with your hands and arms. Try cubes, pyramids and other shapes.

Lesson Notes

1-3. This paper ring should rest level around the middle of the sun. If it tilts, fit the canning ring over the top as a template and reposition the straw "beads."

4. This ecliptic "collar" represents neither a ring of matter (such as the rings of Saturn) nor the sun's equator. Rather, it defines the orientation of the earth-sun plane, an imaginary 2-dimensional surface swept by a line connecting the center of the revolving earth to the center of the sun.

6. Think of the ecliptic circle as a string of 365 dates divided into 12 monthly segments. The *latitude* of each date on this circular calendar defines where the noon sun appears to pass straight overhead, circling through all points west, to the next day's date.

Answers

4a. Exactly 1/2 of the earth extends above the ecliptic plane. Exactly 1/2 of the sun extends beneath it.

4b. No (not if you live north of the Tropic of Cancer or south of the Tropic of Capricorn). All parts of the continental U.S. and Canada, for example, remain above the ecliptic plane in all seasons.

4c. Winter, spring, summer, fall.

4d. Winter solstice, spring equinox, summer solstice, fall equinox.

5a. Where the ecliptic ring touches the globe, the sun appears to cross your zenith (pass straight overhead, at an altitude of 90°).

5b. During summer solstice the sun appears straight overhead on the Tropic of Cancer.

5c. First day of season: Sun appears at zenith:
    summer solstice   on the Tropic of Cancer
    fall equinox   on the equator
    winter solstice   on the Tropic of Capricorn
    spring equinox   on the equator

6. MAR: from 10° S to 5° N. (about 15°)
   APR: from 5° N to 16° N. (about 11°)
   MAY: from 16° N to 22° N. (about 6°)
   JUN: from 22° N to 23.5° N. (about 1 or 2°)

7. The ecliptic plane is represented both as a *ring* surrounding the sun, and as a *circle* drawn on the globe. Encourage students to use *both* forms in their oral explanations. Again, the only written response required in this step is for students to report how many concepts they successfully explained.

An imaginary string connecting the centers of the sun and earth sweeps along the ecliptic plane as the earth revolves: This string defines a circle larger than the model's ecliptic *ring,* and lying in the same plane.

Use your quadrant to show that your model earth's axis tilts 23.5 ° from vertical, the same as the real earth: When the quadrant is held level with earth's equator, the washer swings about 23.5° from the horizontal plane of the ecliptic *ring.*

On the equator in March, as the sun culminates overhead each day, it appears to shift from south to north: The noon sun in early March lies just south of the equator on the ecliptic circle. It crosses at spring equinox, moving north of the equator in late March. (or) The ecliptic *ring* touches the equator at spring equinox. Earlier in March it touches south of the equator, later, north of it.

On the equator in September, as the sun culminates overhead each day, it appears to shift from north to south: Fall equinox happens at the intersection of earth's ecliptic *circle* with the equator. Earlier in September the noon sun is north of this intersection on the ecliptic circle. Later it is south of the equator. (or) The ecliptic *ring* touches the equator at fall equinox. Earlier in September it touches north of the equator, later south of it.

In December the sun stops its apparent southward motion and begins moving north: Earth's ecliptic *circle* reaches its most southerly point at December's winter solstice. (or) As earth reaches winter solstice in December, the ecliptic *ring* touches its most southerly latitude at the Tropic of Capricorn.

The sun never culminates straight overhead north of the Tropic of Cancer or south of the Tropic of Capricorn: Earth's ecliptic *circle* is confined between these two tropics. (or) The ecliptic *ring* brushes all points between these tropics, but no points beyond them.

On the equator, my noontime shadow might point north or south, or straight underfoot: The ecliptic *ring* (or *circle*) touches both sides of the equator. When the culminating noon sun is to the south, overhead, or north of the equator, my shadow points directly opposite: north, straight underfoot, or south respectively.

When any month on the ecliptic ring touches the same month on the ecliptic circle, both the ring and the whole circle lie in the same plane: The ecliptic tilts into horizontal alignment with the ecliptic ring wherever matching dates touch.

Use your quadrant to show that the sun never sets north of the arctic circle on the day of summer solstice: Rest the quadrant on both spheres in summer solstice position to model an incoming parallel sun ray. It touches the "top" of the arctic circle at midnight. The midnight sun dips to meet the horizon at this point, then sweeps back up. (or) The north pole leans directly toward the sun so that *all* points inside the arctic circle tilt into sunlight.

Use your quadrant to show that the sun never rises north of the arctic circle on the day of winter solstice: Rest the quadrant on both spheres in winter solstice position. It touches the "top" of the arctic circle at noon. The noon sun almost rises at this point, then arcs southward. (or) The north pole leans directly away from the sun so that all points inside the arctic circle tilt away from sunlight.

Equinox means "equal nights:" Rest the quadrant on both spheres in spring or fall equinox position. Because the poles tilt sideways with respect to the sun, the circle of twilight between night and day runs directly north-south, through each pole. Every point on the globe turns through 12 hours of night and 12 hours of day.

Solstice means "still sun:" The earth's ecliptic circle reaches its most northerly point at summer solstice and its most southerly point at winter solstice. At these times, the latitude of the culminating noon sun changes very little from one day to the next. It appears to "stand still."

Materials

☐ The model sun and earth.
☐ A regular sized canning ring (optional). The wide-mouth size is too large. Students can level the paper ring by eye, but the task is more difficult.
☐ A straw and scissors.
☐ Masking tape and a rubber band.
☐ The Ecliptic Ring and Concept List cutouts.
☐ The quadrant from earlier activities.

SOLAR TIME / STAR TIME

1. Cut out the <u>Clock Rectangle</u>. Fold it along the center line.

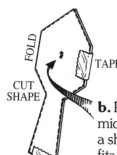

a. Cut out the clock through both layers, leaving the folded side joined. Tape the layers together.

b. Poke a pin through the middle. Drill this hole with a sharp pencil until it *just* fits over a pinhead.

c. Fit the clock, SOLAR side up, over the north pole pin on your globe. Point it at your model sun.

2. Rotate your globe underneath the fixed clock. Count off 24 sun hours (from noon to noon), using your home pin as an "hour hand."

NOON... 1 p.m.... 2 p.m....

YOUR HOME PIN

SUN

a. Suppose it's Monday, 12:00 noon where you live. What is the day and time in England? In Japan?

b. When it's Saturday, 3:00 pm in Western Australia, what is the day and time where you live?

3. Revolve your globe around the sun. Pause at the start of each new season to adjust your home pin and clock to 12:00 noon under the sun (solar noon).

a. Why is it necessary to rotate the earth through an extra quarter turn at each season?

REMEMBER: Keep your paper clips aligned.

...NOW TO SPRING EQUINOX.

b. How many extra quarter turns does the earth make in one year to "keep up" with the sun?

4. As the earth rotates 1 full circle on its axis, it *also* revolves about 1° around the sun.

SUN

1° OF REVOLUTION

EARTH

EARTH'S ORBIT

a. How many degrees, then, does the earth rotate from your solar noon today to your solar noon tomorrow?

b. In 1 full week, the earth rotates 7 full circles *plus* how many degrees more?

c. In 1 full year the earth rotates about 360 full circles *plus* how many degrees more?

5. Pick a point on a *distant* object as far away as you can see (outside a window?). Think of this point as a "star."

a. Turn your SOLAR clock over so the STAR side faces up. Point it at your distant "star."

b. Rotate your globe underneath the fixed clock. Count off 24 star hours (from star noon to star noon) using your home pin as an "hour hand."

"STAR" NOON... 1 p.m.... 2 p.m....

c. How is star noon different from solar noon?

6. Revolve your globe around the sun again. Pause at the start of each new season to adjust your home pin and clock (if necessary) to 12:00 noon under the star (star noon).

a. How many degrees did you rotate the earth at each location?

"STAR"

b. Why is no extra turning required to "keep up" with the star?

c. Which takes more time to complete, a sun day (*solar day*) or a star day (*sidereal day*)? Explain.

EARTH ORBIT

SUN

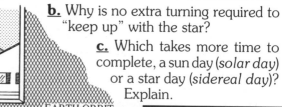

KEEP PAPER CLIPS ALIGNED

7. Cut out and complete the <u>Concept List</u> for this activity. How many concepts did you and your friend understand?

Objective

To recognize that the earth turns through 361° from one solar noon the next. To distinguish between solar time and star time.

Supporting Concepts

✪ Find the international date line on your globe. Places to the east of this line (Hawaiian Islands) are a day behind places to the west (New Zealand).

✪ The sun is a star at the center of our planetary system. Most of the stars we see in the night sky are millions of times farther away.

✪ As you move, close objects appear to shift faster than distant objects. (This is called parallax.)

• Ride down the road in a moving car: telephone poles appear to whiz by, relative to slowly shifting mountain peaks.

• Ride around the sun on the orbiting earth: the sun appears to shift, relative to background stars that appear fixed.

Lesson Notes

2. You may need to demonstrate this procedure to younger students: point the clock at the model sun. Hold it in this position with one hand while turning the earth beneath it with the other.

5. If you have no outside windows, a "star" on the opposite side of the room is far enough away.

Answers

2a. Here is a sample answer for where we live:

Portland, Oregon = Monday, 12:00 noon
England = Monday, 8:00 pm
Japan = Tuesday, 6:00 am

(Local times may vary an hour or so because of time zone variations and daylight savings time.)

2b. Answer depends on your locality.

3a. Each season (1/4 year) the earth revolves a quarter circle around the sun. The home pin and clock must also be rotated an extra quarter turn to keep the eastward shifting sun at its culminating noon position.

3b. The earth rotates through 4 extra quarter turns (360° per year) to catch up with the eastward drifting noon sun.

4a. From solar noon to solar noon the earth rotates 361°.

4b. In 1 full week the earth rotates 7 full circles plus 7° more.

4c. In 1 full year the earth rotates roughly 360 full circles plus 360° more. (The earth really turns 365.26 times per year, plus one turn more. We account for the extra .26 day with a "leap year" every fourth year.)

5c. *Star* noon is when your *"star"* culminates, reaching its highest position in the sky. *Solar* noon happens when the sun culminates.

6a. No extra rotation was necessary to keep the distant star culminating above its home pin.

6b. The star is located so very far away that earth's revolution around the sun causes no apparent shift in the star's position. Throughout the year, each 360° rotation of the earth returns this star to the same culminating position as before.

6c. A solar day takes more time to complete than a sidereal day because it requires 1 extra degree of earth rotation to turn it to an identical position relative to the sun.

7. Students should report how many of the following concepts they successfully explained using their models.

When it's high noon today in Florida, it's 5 pm this afternoon in England, and 3 am tomorrow in Japan: Point the solar clock to Florida and hold it there. Then 5 pm on the clock points toward England, and 3 am points toward Japan. Japan is 1 day ahead because it is west of the international date line.

Central Europe and South Africa share the same time of day even though they are separated by about 10,000 miles: Night and day move west to east across the globe. Distant areas that lie on the same north-south longitude have the same local time.

Set your watch ahead when flying east, back when flying west: When flying east you travel *faster* through the solar day than the rotating earth does. When flying west you move *more slowly* through the solar day than the rotating earth does.

A jet pilot, flying west along the equator at a speed of just over 1,000 miles per hour, observes that the sun remains fixed overhead throughout the 6 hour flight: The pilot is flying west at the same speed that a point on the equator turns east. This creates a "treadmill effect" where the plane maintains a fixed position between the rotating earth below and the sun above.

The earth rotates about 361° each solar day: Earth revolves counterclockwise around the sun. This motion makes the sun appear to drift eastward about 1° during each daily rotation. After rotating 360° from the last solar noon, the earth must rotate 1° more to return the sun to its culmination point.

The earth rotates 360° from one culmination of a reference star to the next: Earth's rotation around the sun causes no apparent shift in the reference star's position because the star is so far away. One 360° rotation of the earth, therefore, returns that star to its same culminating position in the sky.

Your "reference star" and model sun culminate together just once a year. When this happens, solar time agrees with sidereal time: There is only one point in earth's orbit around the sun where both the reference star and the sun line up on the same side of the earth. At this point, solar time and sidereal time briefly agree.

If you could observe your reference star during the same solar time each day, it would appear to drift slowly west relative to the sun: The earth rotates 360° in an easterly direction to return a reference star to its same sky position, then 1° more to return the sun to its same sky position. With each passing day, the earth turns eastward another 1° beyond the reference star, shifting it west relative to the sun.

If the earth revolved but did not rotate, the sun would appear to drift eastward through a day and night that lasted one year: If the earth revolves counterclockwise without rotating, the sun appears to slowly rise in the *west* as it begins 6 months of day. Halfway through its yearly orbit, the sun slowly sets in the *east* to begin 6 months of night.

If the earth rotated counterclockwise just once each year, the sun would appear to stand perfectly still in the sky all year long: Rotating the earth just once through its yearly revolution around the sun keeps the same side of the earth facing the sun at all times. (In the next activity, your students will learn that this is how the moon moves around the earth.)

There are 1440 star minutes in 1 star day. Every 4 star minutes the earth rotates 1°: Twenty four star hours multiplied by 60 minutes per hour equals 1440 star minutes. Dividing by 360° gives 4 star minutes per degree.

A sidereal day lasts about 23 solar hours and 56 solar minutes: The earth rotates through a 361° solar day in 24 solar hours. A sidereal day lasts 1° less, that is, 24 solar hours minus 4 star minutes. In solar time this equals *about* 23 hours and 56 minutes.

Materials

☐ The Clock Rectangle cutout.
☐ Scissors.
☐ Clear tape.
☐ A pin.
☐ A pencil sharpener.
☐ The model earth and sun.
☐ The Concept List cutout.

MOON MOTION

1. Stick a tape "collar" around the mouth of a bottle. Fringe it as before, and fan it outward.
a. Get your moon phase model. Hold it up to strong light so the black center spot shows through the bottom of the paper plate.

FRINGED TAPE

b. Stick the mouth of the bottle under this spot.

2. Line up these 3 models from left to right across a table as shown. Turn the earth and sun to summer solstice position.

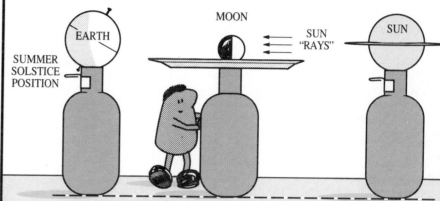

MOON
EARTH
SUN "RAYS"
SUN
SUMMER SOLSTICE POSITION

a. Turn the plate so its "sun rays" point from sun to earth. What moon phase do you now see from your model earth?
b. Revolve your model moon counterclockwise around the earth, so the sun rays always come from the sun. List each moon phase (in order) that you see from earth.

3. Revolve the *moon* around the earth to first quarter. Rotate the earth through a 24 hour cycle at this position.

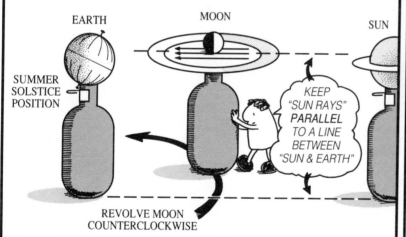

EARTH
MOON
SUN
SUMMER SOLSTICE POSITION

KEEP "SUN RAYS" PARALLEL TO A LINE BETWEEN "SUN & EARTH"

REVOLVE MOON COUNTERCLOCKWISE

a. When and where (from your home pin) does the *first quarter* moon rise? Where does it culminate? Where does it set?
b. Answer this question again for a *full* moon.
c. Answer this question again for a *third quarter* moon.

4. Model the moon with a ball of clay. Mark a "face" on it to represent the "man in the moon."

a. As the moon revolves around the earth, we see *only* its face, *never* the "back of the head." Demonstrate this.
b. Does the moon rotate as it revolves? If so, how much?
c. Can your paper plate model show this rotation? Why?

5. Review your moon observations from activity 3.
a. In which direction did the moon move during ONE NIGHT? How can you interpret this motion with your models?
b. How can you use your models to interpret the moon's motion over ONE WEEK.

6. Complete the Concept List for this activity. How many concepts did you and your friend understand?

Objective

To model the moon's *apparent* westward motion and *actual* eastward motion around the globe. To correlate the moon's phases with its positions relative to the earth and sun.

Supporting Concepts

⊙ The moon circles the earth about once a month.

⊙ "Quarter moons" refer not to apparent size or shape, but to *position* in the monthly moon phase cycle. The first quarter moon is 1/4 cycle from new moon to new moon around the paper plate; the third quarter is 3/4 of the way around. (Because the first and third quarter moons *look* like "half" moons, they are sometimes called this.)

Lesson Notes

2b. Because the sun's rays travel over such a huge distance, they strike the earth and moon virtually parallel, as modeled. Students should rev**o**lve this Ping Pong moon around the globe, so these "sun rays" always point in the same direction, *parallel* to an imaginary line between the earth and moon models.

Answers

2a. The new moon phase.

2b. New moon, waxing crescent, first quarter, waxing gibbous, full moon, waning gibbous, third quarter, waning crescent. Then the cycle repeats.

3a. The first quarter moon rises in the east at noon. It culminates toward the south at sunset. It sets in the west at midnight.

3b. The full moon rises in the east at sunset. It culminates toward the south at midnight. It sets in the west at sunrise.

3c. The third quarter moon rises in the east at midnight. It culminates toward the south at sunrise. It sets in the west at noon.

4b. Ye̅s. To keep the same side of the moon facing earth, it rotates once on its axis each time it completes one revolution around the earth. (Tidal forces between the earth and moon have acted over eons of time to slow the spinning moon to this "lock-step" rotation. An observer on the moon's face would always see our beautiful earth in the same part of its inky black sky. Only earth's phases would change, waxing and waning like the moon.)

4c. No. The Ping Pong ball cannot be rotated because its "dark" side, covered with black tape, must remain turned away from the sun.

5a. Observing at successive hours over the course of ONE NIGHT, the moon appeared to move from east to west. This apparent motion is modeled by rotating the globe counterclockwise, from west to east.

5b. Observing at the same hour over the course of ONE WEEK, the moon appeared to move from west to east. This actual motion is modeled by revolving the moon counter-clockwise, from west to east around the earth.

6. Students should report how many of the following concepts they successfully explained using their models.

A full moon always culminates at midnight: When the sun shines on the moon with the earth in between, we see all of the moon's face fully illuminated. This full moon is visible everywhere on the dark side of the earth, and highest in the southern sky at midnight.

A new moon cannot be seen unless it eclipses the sun: When the moon rev**o**lves between the earth and sun, its shadowed side faces us. We see nothing of its dark form near the brilliant sun, unless this new moon actually eclipses the sun, appearing to cross its face.

Quarter moons always culminate at sunrise or sunset: When the sun shines on the moon from the "side," we see equal portions of its surface in light and shadow. This first or third quarter moon is visible on earth only between noon (if viewing conditions are right) and midnight. It culminates in the southern sky halfway between these limiting times, illuminated from the "side" by a sun that sets or rises on our horizon.

Thin crescent moons are seen just after sunset or just before sunrise: The moon exposes a thin crescent of its illuminated side when it is just outside of its new-moon alignment between the earth and sun. This crescent is outshined by the sun and rendered quite invisible until, at sunrise and sunset, the sun's brilliance is shielded by the horizon.

A waxing moon is a more familiar sight than a waning moon: A waxing moon is visible during evening waking hours as the earth rotates away from sunset. A waning moon is visible during morning sleeping hours as the earth rotates into sunrise.

A new moon can eclipse the sun, but a full moon never can: The moon eclipses the sun as it rev**o**lves directly between the earth and sun to cast its shadow on the earth. This shadow is directed toward the earth only at new moon. It points away at full moon.

The earth can eclipse a full moon but never a new moon: The earth's shadow always points directly away from the sun. Only a full moon can pass through this shadow.

During one night, the moon appears to move from east to west across the sky: The earth rotates counterclockwise from west to east, apparently moving the moon from east to west.

Over many nights, the moon appears to move from west to east across the sky: Earth's rotation makes the moon appear to move westward even though it actually rev**o**lves eastward around the earth. Over several nights, the moon's eastward revolution carries it far enough to easily recognize.

The same side of the moon always faces the earth: Draw a moon face on a lump of clay. To keep this face turned toward the earth at all times you must rotate it through one complete turn as it revolves one full circle around the globe.

A winter full moon culminates higher in the sky than a summer full moon: During winter, the northern hemisphere tilts away from the noon sun and thus toward the midnight full moon. During summer, opposite conditions prevail.

*The moon takes about 27 days to rev**o**lve full circle (360°) around the earth. It takes about 2 additional days to return to the same phase:* During the moon's 27 day revolution around earth, the earth has time to revolve through a significant part of its orbit around the sun. This advances the sun eastward in the sky, requiring the moon to revolve 2 additional days eastward to again reach its starting point (relative to the sun) in the moon phase cycle.

Extension

Improve your model so you can accurately represent the moon's rotational motion on its axis. (Carefully peel back the cap of black tape, turning it sticky-side-out. Pencil in basic lunar features on the ball, using photos or moon maps. [Or draw a face on the moon.] Push the finished moon back into its black cap. Recenter the edge of this sticky cap on the paper plate. Operate your model as before. In addition, rotate the moon inside its cap so its same face continuously turns toward earth.)

Materials

☐ Masking tape and scissors.
☐ A glass beverage bottle of similar height to your model earth and sun supports.
☐ The moon phase model from activity 13.
☐ The earth and moon models from previous activities. Remove the clock hanging at the north pole.
☐ Clay.
☐ The Concept List for this activity.

LUNAR CALENDAR

A. *Get both* <u>Lunar Cycle</u> *cutouts, plus calendar information for the next 12 months. Use pencil to complete steps 1-11* **in order:**

2. Write the weekday and date of the *next full moon* here.

5. Count calendar days between the full moon and *next* new moon. Shade dashed sections as in step 3.

3. Count the *calendar* days from new moon to full moon. (Call the new moon "zero.")

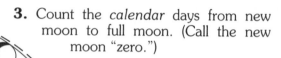

- If you count *14* days, darken *2* dashed sections.
- If you count *15* days, darken *1* dashed section.
- If you count *16* days, darken *no* sections.

6. Neatly complete the last half of the moon cycle, up to (but not including) the next new moon. If the month changes, separate the dates with a *very* heavy line.

4. Neatly fill in all dates and weekdays between these moons. (Keep all numbers and letters upright as shown.) If the month changes, draw a *very* heavy line where the new one begins.

7. Write the ending month here.

8. Cut out this lunar cycle on the dashed line. Tape its tiny "x" to the inside end of a "ripple" on a paper plate.

START HERE: 1. Go to the Lunar Cycle numbered ONE. Notice how the date, weekday, month and year have been written for a *new moon*.

9. Complete the remaining lunar cycles numbered 2-12 *in order:*
   a. Start each new cycle at the next new moon.
   b. Tape each cycle 30° to the *left* of the last by its tiny "x". (If a plate has 72 ripples, each one equals 5°.)

WHEN ALL 12 CYCLES ARE FILLED IN, GO ON TO STEP 10!

10. Cut out the <u>Seasons Circle</u>. Fix it to the center of the plate with rolled tape so "summer solstice" on this circle lies next to the lunar cycle containing the summer solstice (on or near June 21).

11. Pull your current moon cycle in front of the others so the whole circle shows. Cut out the <u>Moon Marker</u>. Slip it behind today's date with the moon tab folded up.

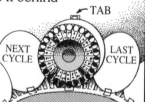

B. *Examine your lunar calendar to answer each question:*
<u>**1.**</u> How long, on average, is a lunar month?
<u>**2.**</u> Do 12 lunar months equal a solar year? Explain.
<u>**3.**</u> How does earth's axis currently tilt relative to the sun? What season are you now in?
4. Each moon watcher is standing at about 35°N latitude, and rotating one line of longitude per hour. Mark *your* current position on this month's globe with a small "x."
<u>**5.**</u> Where is the moon *right now* relative to your home? What phase is it in?
<u>**6.**</u> Define these cycles in terms of the earth, sun and moon: year, season, month, week, day.

C. *Take your calendar home.*

Track your daily motion through the great wheels of time, all year long!

Objective

To construct a lunar calendar. To summarize basic earth and moon motions in a carry-home model to use all year long.

Supporting Concepts

Call attention to these important details on the Lunar Cycle cutouts:

• How far does a moon watcher on earth rotate in 24 hours? (About full circle, 361°.) How far does the moon *actually* revolve in this much time? (From one moon picture to the next. It *appears* to move nearly full circle because the earth is rotating.)

• What makes each moon cycle unlike all the rest? (The earth in the middle has a different tilt toward the sun.)

• Notice that all the moon watchers stand on the same circle. If this circle represents 35° N latitude, where is your home latitude? (Somewhat *inside* this circle if your local latitude is greater than 35° N; *outside* if it is less.)

• Clock times surrounding each earth are in standard time. Subtract 1 hour to convert from daylight savings time. Where is the sun at 1 p.m. daylight savings time? (In the culminating 12-noon position.)

• These model earths are divided into 24 hours of longitude. Starting at your nearest local hour, trace your movement on this earth through a 24 hour rotation. (Move counterclockwise, 1 line per hour.)

• Suppose the moon is full:
 Where does the *midnight* watcher see it? (Culminating south.)
 Where does the *6 am* watcher see it? (Setting, western horizon.)
 Where does the *6 pm* watcher see it? (Rising, eastern horizon.)
 Where does the *noon* watcher see it? (She doesn't. It's on the opposite side of the earth.)

Preparation

<u>Prepare the Lunar Cycle cutouts:</u>

(1) List the numbers 1-13 on paper. Starting with the calendar date of your most recently completed new moon, list the dates of 13 consecutive *new* moons next to these numbers. This time span (about 355 days) will likely carry you forward into the next calendar year. If you can't find next year's calendar with moon phases noted, start with an "older" new moon, one that occurred *before* January 26.

(2) Cross out the 13th new moon. (Your calendar will end 1 day short of this date.) Pick the beginning of any new season (either a solstice or equinox date) that falls within the first 3 weeks of a moon cycle, but *not* in the last week (fourth quarter). Insert the name (not date) of this season change into your list *between* the new moons in which it occurs. Circle the line number directly *above* it.

(3) Make 1 master photocopy of Lunar Cycles A and B. Notice that each sheet has 6 boxes; find the one labeled with the season change you just picked. Copy the number you just circled into this lunar cycle box. From here, number all remaining boxes in reading sequence: from left to right down each sheet (...9, 10, 11, 12, 1, 2, 3, ...).

(4) Write the date, weekday, month and year of the first new moon on your list at the bottom of the lunar cycle you numbered 1. Follow the pattern illustrated on the *worksheet* sample dated "Tuesday, November 24, 1992." You can provide extra help for younger students by filling in the remaining new moons (and even full moons and quarter moons). However, the more students do for themselves, the more pride they will take in their finished calendars.

(5) Photocopy these masters, 1 set of sheets A and B per student.

<u>Prepare reference calendars:</u>

(1) Find the three small calendars printed on the back of your current check register. Mark your starting new moon date with a *dark* circle in the correct year. (It's OK to obscure the number.)

(2) Mark the next *full* moon with an *open* circle. Continue marking new and full moons through all 12 lunar cycles with dark and open circles.

(3) These calendars are now ready to photocopy, one per student. Enlarge them if your copy machine has that capability.

Lesson Notes

A. Students should work in pencil to make errors easy to correct. Be sure they start with step 1, on the right side of the page.

3. If there are 15 days between the new and full moon, you must darken just 1 of 2 possible spaces. This is a random choice, unless

you are using a calendar with quarter moon dates also identified. In this case, fill in the quarter moon date first, then darken the dashed section in the quarter cycle that requires only 7 days.

7. Nearly all lunar cycles begin in one month and end in the next. If the new moon occurs during the first three days of a new month, however, the cycle may start and end in that same month.

8-9. It takes about an hour to neatly fill in all 12 lunar cycles. Depending on your teaching strategy, it may be a good idea to ask students to fill in all 12 cycles first, then begin cutting and taping at the start of another new period.

10. If your calendar has no summer solstice, align the circle with winter solstice instead (on or near December 21). This happens only if you start your calendar soon after June 21, relegating summer solstice to the 11-12 day gap between 12 lunar cycles and 1 solar year.

Remember that these solstice and equinox labels only roughly point to their corresponding dates. The earth, after all, revolves continuously around the sun. It does not jump from cycle to cycle.

11. While the earth is drawn from an outside perspective, looking "down" on the north pole, the moons around it are pictured from an earthbound perspective. These different orientations are reconciled by the small moon marker. Its vertical tab suggests that each moon watcher sees these moons projected vertically. To see the moon as it actually appears in the sky, we must stand in the shoes of a tiny moon watcher and mentally slide each moon image up from the elliptic plane into the vertical plane of the marker.

B. We have drawn the 4 moon watchers on each globe so they appear to stand on a circle at about 35° N latitude. To roughly estimate your own latitude on the earth corresponding to your current date, find that point where your hour of sunrise or sunset (in standard time) intersects the edge of the twilight shadow.

Answers

B1. About half of the lunar months have 30 days (32 spaces, less 2 shaded), the other half 29 days (32 spaces, less 3 shaded). On average, a lunar cycle lasts 29.5 days.

B2. Almost. There is a 11-12 day gap between the calendar's starting date and its ending date. Another 1/3 lunar cycle is required to fully complete the solar year. (This ragged match between whole lunar cycles and a full solar year has challenged calendar makers throughout human history.)

B3. Answers depend on time of year. (The moon marker at today's date shows earth's current position relative to the sun. Its north pole, tilting toward darkness, light, or in between, defines the season.)

B4. The placement of this "x" depends on your locality and time of day. (Los Angeles, Oklahoma City, Memphis and Charlotte are all located near 35° N latitude. Move inside this circle if you live further north, outside if you live further south. Your local time, less 1 hour for daylight savings time, fixes your longitude.)

B5. Answers depend on your calendar date and the time of day. (The moon is currently positioned at today's date. Relative to your home position marked by an "x" it might be overhead, on the horizon, or "behind" the earth. To estimate your horizon positions, count 6 hours east and 6 hours west from your local standard time.)

B6. Year: One earth revolution of the sun.
 Season: A quarter revolution of the sun.
 Month: One moon revolution of the earth.
 Week: One quarter of a moon cycle.
 Day: A 361° rotation of earth from solar noon to solar noon.

Materials

☐ Two Lunar Cycle cutouts (labeled A and B) per student. Prepare these as detailed in "Preparations" above.

☐ Calendar information. Supply actual calendars showing moon phases if just one or two students will use them at one time. Otherwise, supply photocopies of your check register calendars. See "Preparations" above.

☐ Scissors and clear tape.

☐ A 9 inch (22.9 cm) paper plate. Generic brands often have 72 ripples per 360° circle, so that each ripple occupies 5°. This allows accurate placement of the 12 lunar cycles at 30° intervals by taping them 6 ripples apart. If your plates are not so conveniently marked, supply students with rulers and protractors.

☐ The Seasons Circle and Moon Marker cutout sheet.

SUPPLEMENTARY
CUTOUTS
(13 PAGES)

COMPASS CIRCLE
ACTIVITY 1

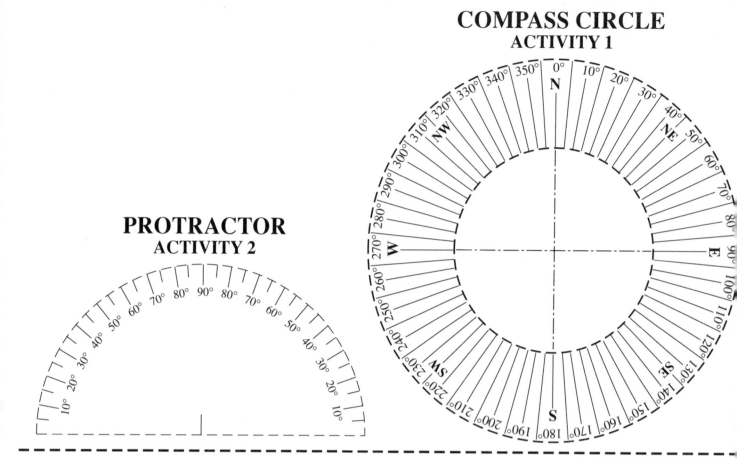

PROTRACTOR
ACTIVITY 2

ANGLES OF ALTITUDE
ACTIVITY 2

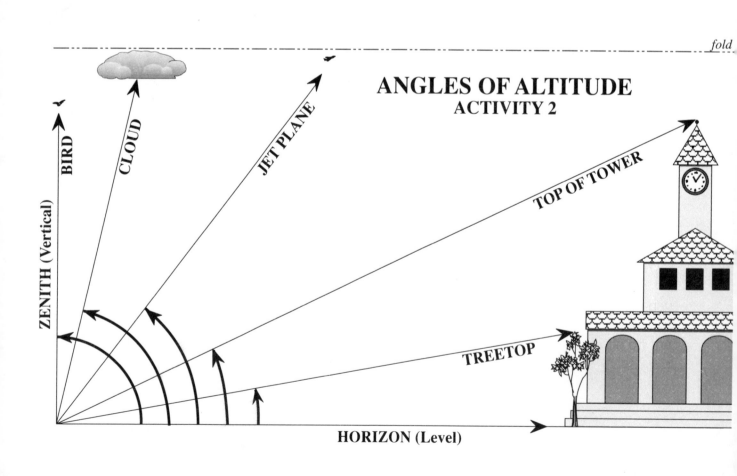

fold

BIRD
CLOUD
JET PLANE
TOP OF TOWER
TREETOP
ZENITH (Vertical)
HORIZON (Level)

QUADRANT
ACTIVITY 2

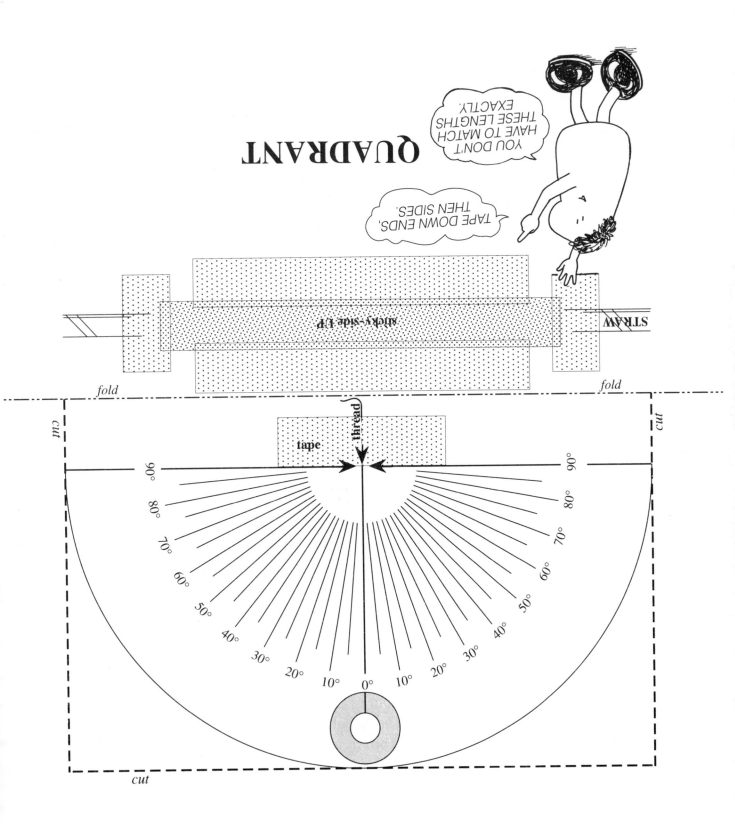

ACTIVITY 3
CIRCLE GRAPHS

CORNER POINTS

Observations
For **ONE WEEK**

Observations
For **ONE NIGHT**

ACTIVITY 4
CIRCLE GRAPH

Solar Survey

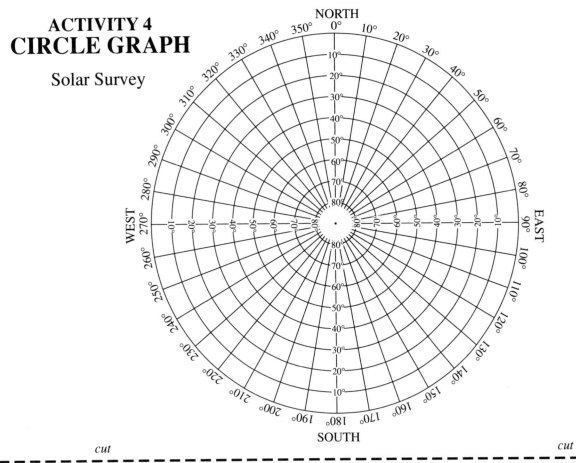

cut   cut

ACTIVITY 5
CIRCLE GRAPH

Shadow Track

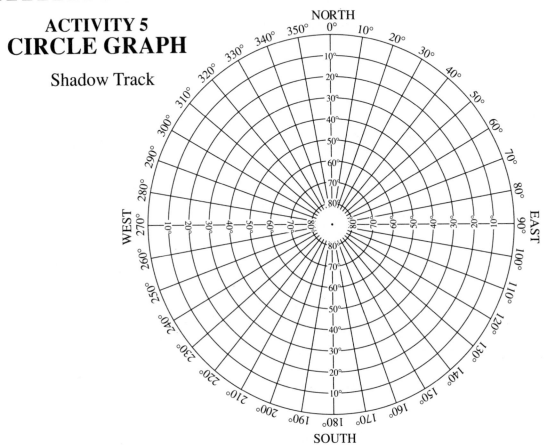

ACTIVITY 5
SUNDIAL CIRCLE

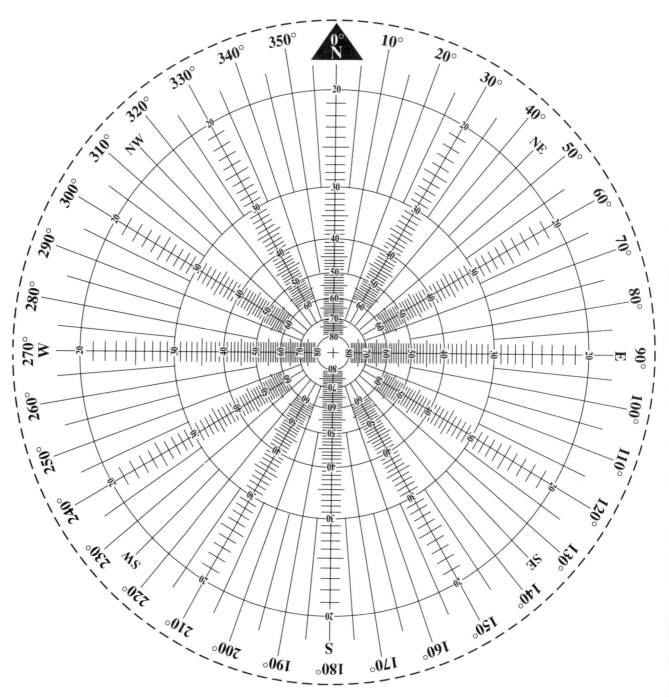

ACTIVITY 10 SUN CIRCLES

ACTIVITY 11 EARTH CIRCLES

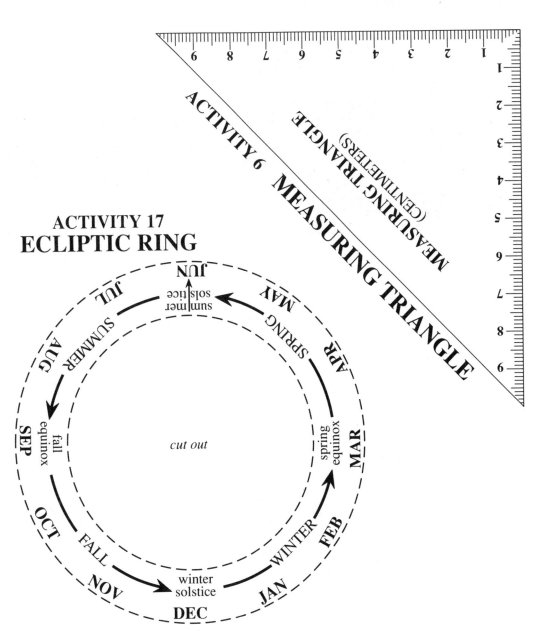

ACTIVITY 17
ECLIPTIC RING

MEASURING TRIANGLE

ACTIVITY 6 MEASURING TRIANGLE
(CENTIMETERS)

JUN
JUL
AUG
SEP
OCT
NOV
DEC
JAN
FEB
MAR
APR
MAY

SUMMER
FALL
WINTER
SPRING

summer solstice
fall equinox
winter solstice
spring equinox

cut out

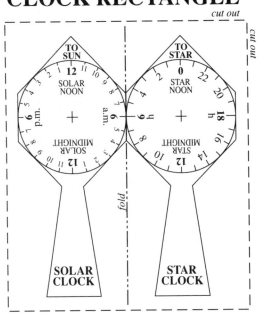

ACTIVITY 18
CLOCK RECTANGLE

cut out

TO SUN
12
SOLAR NOON
p.m.
a.m.
SOLAR MIDNIGHT
12

TO STAR
0
STAR NOON
h
STAR MIDNIGHT
12

fold

cut out

SOLAR CLOCK

STAR CLOCK

MOON PHASE TABS
Activity 13

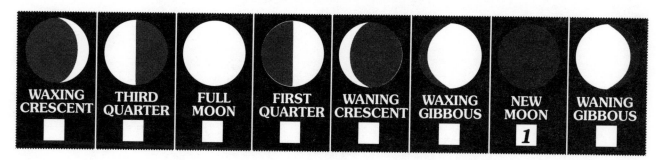

WAXING CRESCENT	THIRD QUARTER	FULL MOON	FIRST QUARTER	WANING CRESCENT	WAXING GIBBOUS	NEW MOON	WANING GIBBOUS
☐	☐	☐	☐	☐	☐	1	☐

SUNLIGHT CIRCLE
Activity 13

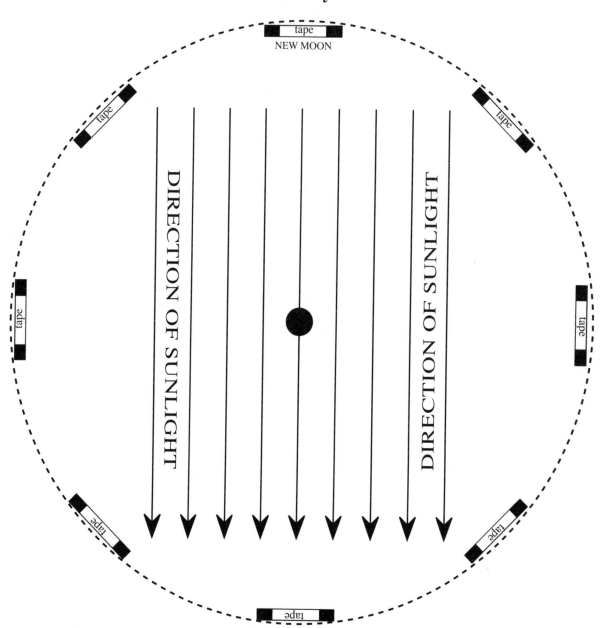

ACTIVITY 14
GLOBE GORES

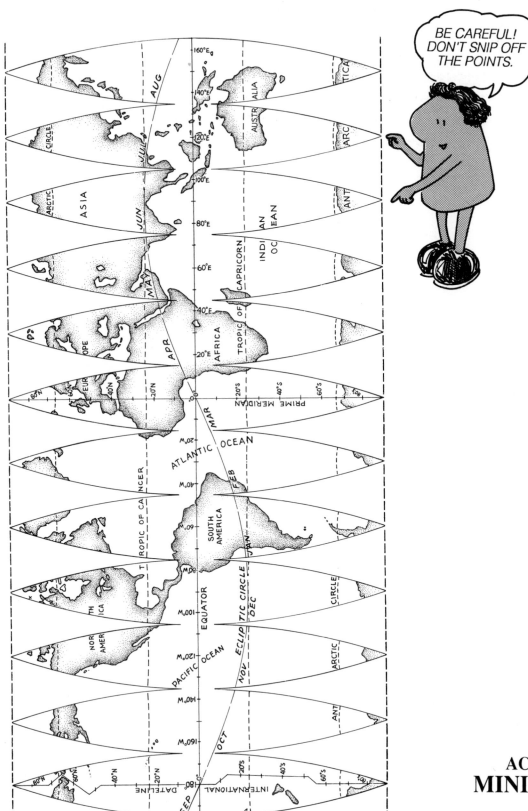

Activity 16
CONCEPT LIST

- [ ] It gets light and dark once a day.
- [ ] The sun appears to rise in the east and set in the west.
- [ ] Each year, there is a warm summer season and a cold winter season.
- [ ] The earth's warm and cold seasons are separated by a milder spring and fall.
- [ ] Summer in one hemisphere is winter in the other hemisphere.
- [ ] Winter and summer happen once in a year.
- [ ] Winter nights are longer than summer nights.
- [ ] Your noontime shadow is shortest in summer, longest in winter.
- [ ] Your shadow at high noon always points north in Canada, south in Southern Australia.
- [ ] Winter nights are much longer near the poles than at the equator.
- [ ] At earth's poles, the sun circles nearly parallel to your horizon, rising and setting just once a year!
- [ ] Twilight near the poles is much longer than twilight near the equator.

Activity 17
CONCEPT LIST

- [ ] An imaginary string connecting the centers of the sun and earth sweeps along the ecliptic plane as the earth revolves.
- [ ] Use your quadrant to show that your model earth's axis tilts 23.5 ° from vertical, the same as the real earth.
- [ ] On the equator in March, as the sun culminates overhead each day, it appears to shift from south to north.
- [ ] On the equator in September, as the sun culminates overhead each day, it appears to shift from north to south.
- [ ] In December the sun stops its apparent southward motion and begins moving north.
- [ ] The sun never culminates straight overhead north of the Tropic of Cancer or south of the Tropic of Capricorn.
- [ ] On the equator, my noontime shadow might point north or south or straight underfoot.
- [ ] When any month on the ecliptic ring touches the same month on the ecliptic circle, both the ring and the whole circle lie in the same plane.
- [ ] Use your quadrant to show that the sun never sets north of the arctic circle on the day of summer solstice.
- [ ] Use your quadrant to show that the sun never rises north of the arctic circle on the day of winter solstice.
- [ ] Equinox means "equal nights."
- [ ] Solstice means "still sun."

Activity 18
CONCEPT LIST

☐ When it's high noon today in Florida, it's 5 pm this afternoon in England, and 3 am tomorrow in Japan.

☐ Central Europe and South Africa share the same time of day even though they are separated by about 10,000 miles.

☐ Set your watch ahead when flying east, back when flying west.

☐ A jet pilot, flying west along the equator at a speed of just over 1,000 miles per hour, observes that the sun remains fixed overhead throughout the 6 hour flight.

☐ The earth rotates about 361° each solar day.

☐ The earth rotates 360° from one culmination of a reference star to the next.

☐ Your "reference star" and model sun culminate together just once a year. When this happens, solar time agrees with sidereal time.

☐ If you could observe your reference star during the same time each solar day, it would appear to drift slowly west relative to the sun.

☐ If the earth revolved but did not rotate, the sun would appear to drift eastward through a day and night that lasted one year.

☐ If the earth rotated counterclockwise just once each year, the sun would appear to stand perfectly still in the sky all year long.

☐ There are 1440 star minutes in 1 star day. Every 4 star minutes the earth rotates 1°.

☐ A sidereal day lasts about 23 solar hours and 56 solar minutes.

- -

Activity 19
CONCEPT LIST

☐ A full moon always culminates at midnight.

☐ A new moon cannot be seen unless it eclipses the sun.

☐ Quarter moons always culminate at sunrise or sunset.

☐ Thin crescent moons are seen just after sunset or just before sunrise.

☐ A waxing moon is a more familiar sight than a waning moon.

☐ A new moon can eclipse the sun, but a full moon never can.

☐ The earth can eclipse a full moon but never a new moon.

☐ During one night, the moon appears to move from east to west across the sky.

☐ Over many nights, the moon appears to move from west to east across the sky.

☐ The same side of the moon always faces the earth.

☐ A winter full moon culminates higher in the sky than a summer full moon.

☐ The moon takes about 27 days to revolve full circle (360°) around the earth. It takes about 2 additional days to return to the same phase.

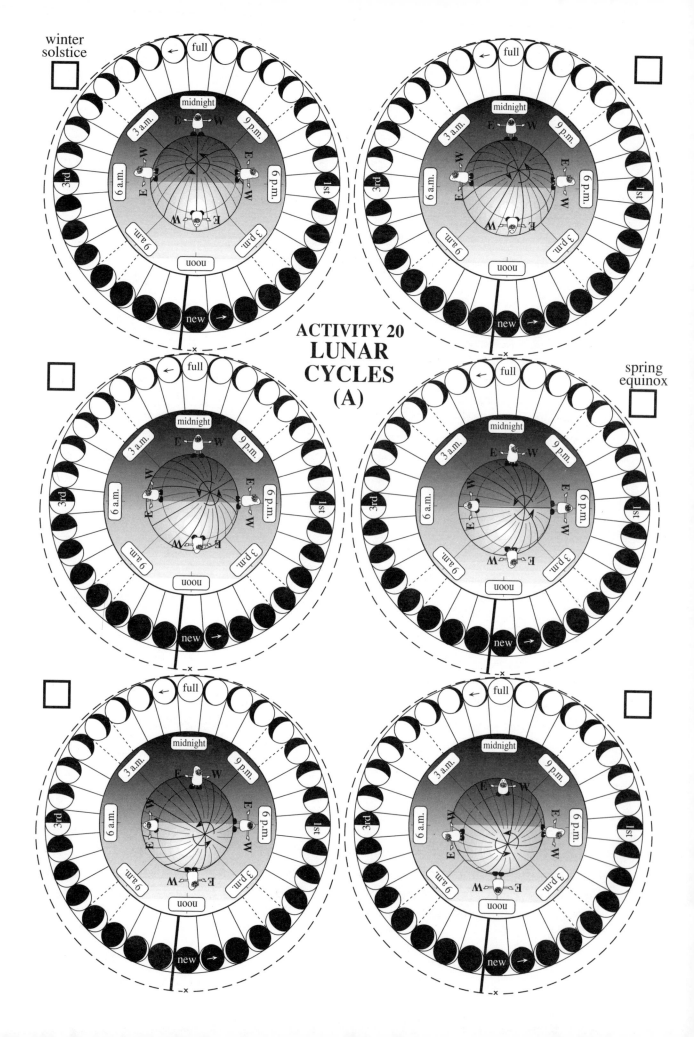

ACTIVITY 20
LUNAR
CYCLES
(A)

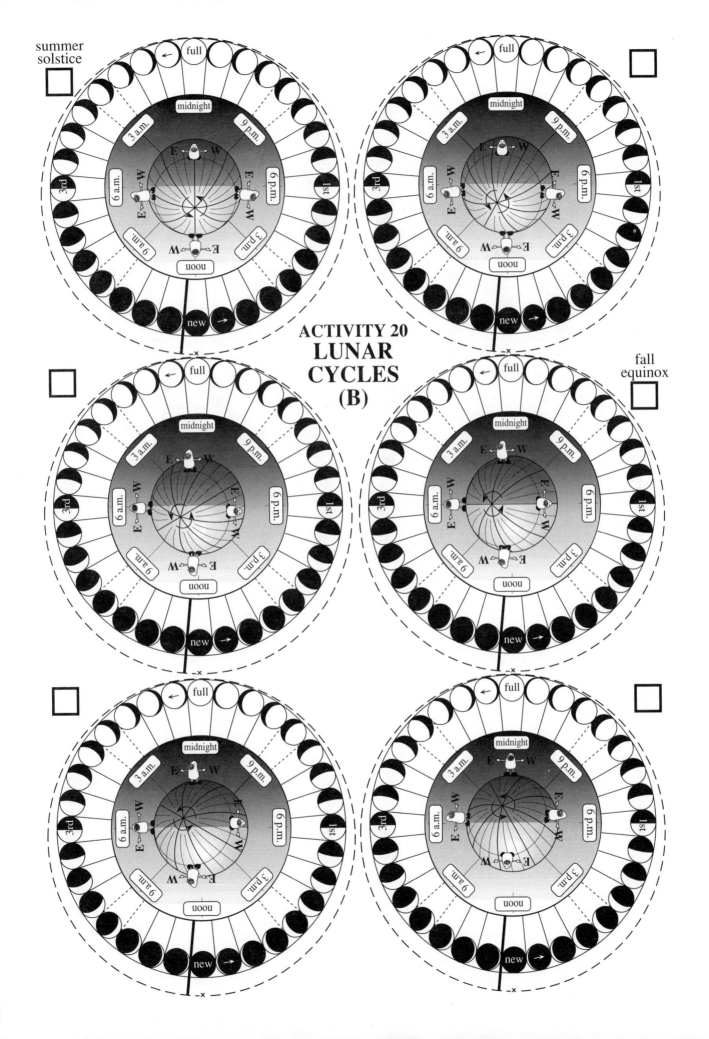

summer
solstice

ACTIVITY 20
LUNAR
CYCLES
(B)

fall
equinox

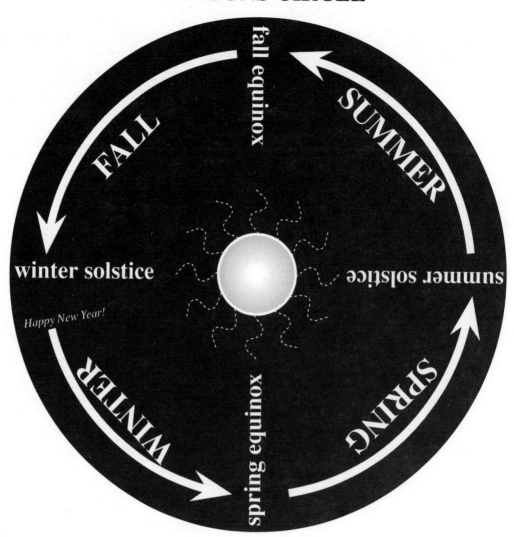

ACTIVITY 20
MOON MARKER